MATH KNOTS
Joy of learning....

My Book

This book belongs to

Cover Design by :
Gowri Vemuri

First Edition :
December , 2019

Author :
Gowri Vemuri

Edited by :
Raksha Pothapragada
Ritvik Pothapragada

Questions: mathknots.help@gmail.com

NOTE : VDOE is neither affiliated nor sponsors or endorses this product.

Dedication

This book is dedicated to:

My Mom, who is my best critic, guide and supporter.

To what I am today, and what I am going to become tomorrow,

is all because of your blessings, unconditional affection and support.

This book is dedicated to the

strongest women of my life ,

my dearest mom

and

to all those moms in this universe.

G.V.

The Virginia Board of Education and Virginia Department of Education (VDOE) have developed the Virginia Assessment Program (VAP) to measure and evaluate students' academic progress in the Standards of Learning (SOLs). The SOLs indicate Virginia's expectations for what students should know and be able to do in the subject areas of reading, writing, mathematics, science, and history/social science.

Students in grades 3-12 to take the Standards of Learning (SOL) assessments each year. Some of the tests are required by all students each year, and others are required only at specific grade levels. Additionally, with the removal of some SOL tests in recent years, the VDOE assigned the responsibility of the creation and administration of alternate, performance - based assessments on local divisions. Student scores from these tests determine a school's and the division's state accreditation and measures progress toward meeting federal targets.

Virginia Standards of Learning (SOL) tests are generally given online unless a student has an identified and documented need to be assessed using paper,pencil format. The test question format is typically multiple choice, and each test contains some technology enhanced items.

GRADE 3-8 STANDARDS OF LEARNING(SOL) TESTS

GRADE 3	GRADE 4	GRADE 5	GRADE 6	GRADE 7	GRADE 8
	VIRGINIA STUDIES				WRITING
MATH	MATH	MATH	MATH	MATH	MATH
READING	READING	READING	READING	READING	READING
		SCIENCE			SCIENCE

NOTE : VDOE is neither affiliated nor sponsors or endorses this product.

END OF COURSE STANDARDS OF LEARNING (SOL) TESTS

GRADE 9	GRADE 10	GRADE 11
ALGEBRA I	GEOMETRY	ALGEBRA II
EARTH SCIENCE	BIOLOGY	CHEMISTRY
WORLD HISTORY I	WORLD HISTORY II	VIRGINIA & US HSTORY
		WORLD GEOGRAPHY
		ENGLISH : READING
		ENGLISH : WRITING

Any Student taking one of the courses listed here is expected to take the corresponding end-of-course SOL test. The grade levels depicted here represent grade level at which students typically participate in these courses.

NOTE : VDOE is neither affiliated nor sponsors or endorses this product.

SOL Test Scoring and Performance Reports:

Standards of Learning assessments in English reading, mathematics, science and history/social science are made up of 35-50 items or questions that measure content knowledge, scientific and mathematical processes, reasoning and critical thinking skills. English writing skills are measured with a two-part assessment that includes multiple-choice items and an essay.

Student performance is graded on a scale of 0-600 with 400 representing the minimum level of acceptable proficiency and 500 representing advanced proficiency. On English reading and mathematics tests, the Board of Education has defined three levels of student achievement: basic, proficient, and advanced, with basic describing progress towards proficiency.

Performance Achievement Levels:

- The achievement levels for grades 3-8 reading and mathematics tests are: *Pass/Advanced*, *Pass/Proficient*, *Fail/Basic*, and *Fail/Below Basic*.

- The achievement levels for science tests, history tests, and End-of-Course (EOC) tests* are: *Pass/Advanced*, *Pass/Proficient*, and *Fail/Does Not Meet*.

- The EOC Writing (2010 SOL) test, EOC Reading (2010 SOL) test, and EOC Algebra II (2009 SOL) test have an achievement level of *Advanced/College Path* in place of the *Pass/Advanced* achievement level.

NOTE : VDOE is neither affiliated nor sponsors or endorses this product.

INDEX

Contents	Page No

FORMULA SHEET

1. Area of a triangle

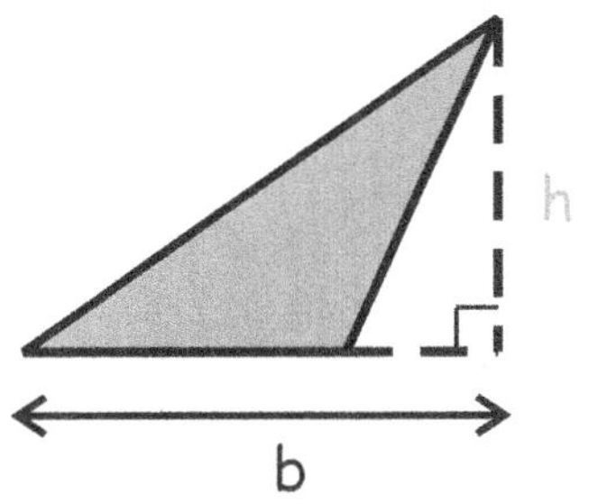

$A = \frac{1}{2} bh$

2. Area of a parellelogram

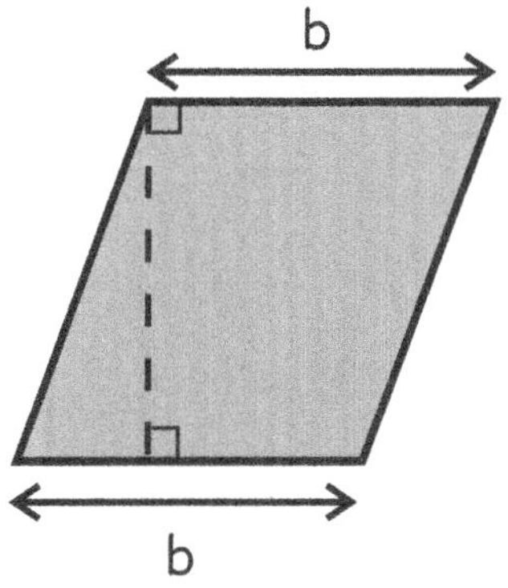

$A = bh$

3. Volume and Surface area of a cuboid

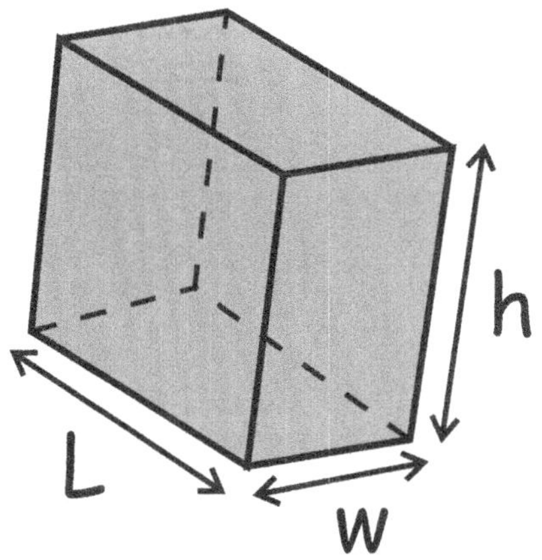

$V = lwh$

$S.A = 2(lw + lh + wh)$

4. Perimeter and Area of a Square

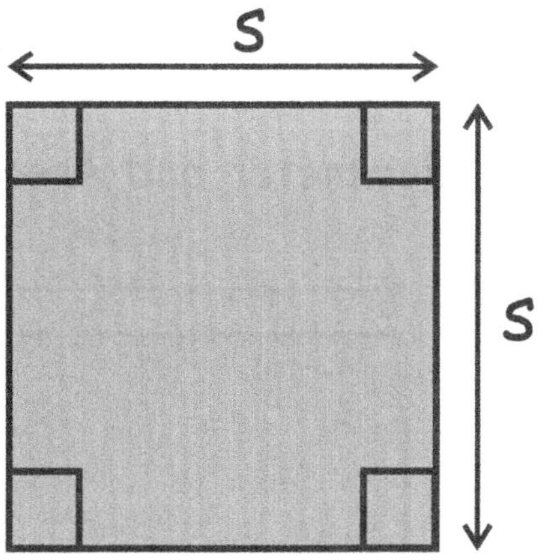

$P = 4s$

$A = s^2$

5. Area of a Trapezium

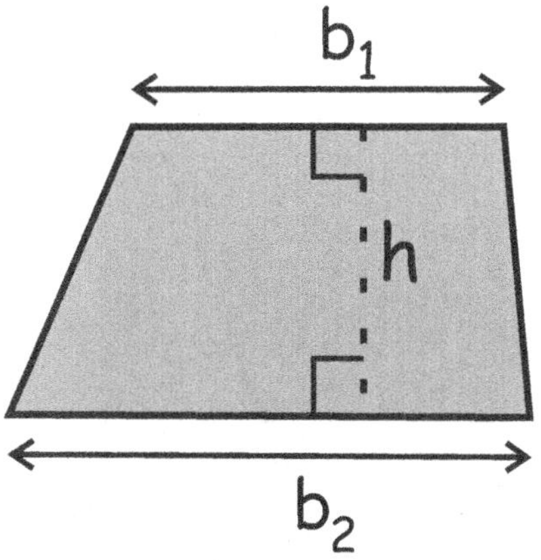

$$A = \frac{1}{2} h(b_1 + b_2)$$

6. Circumference and Area of a Circle

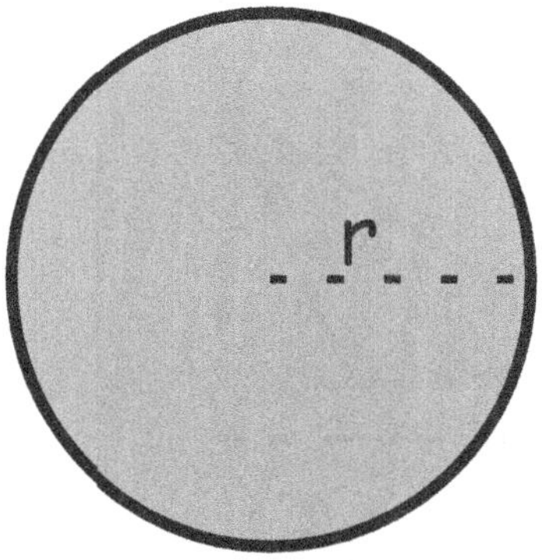

$$c = 2\pi r$$
$$A = \pi r^2$$

pi

$$\pi = 3.14$$
$$\pi = \frac{22}{7}$$

7. Perimeter and Area of a Rectangle

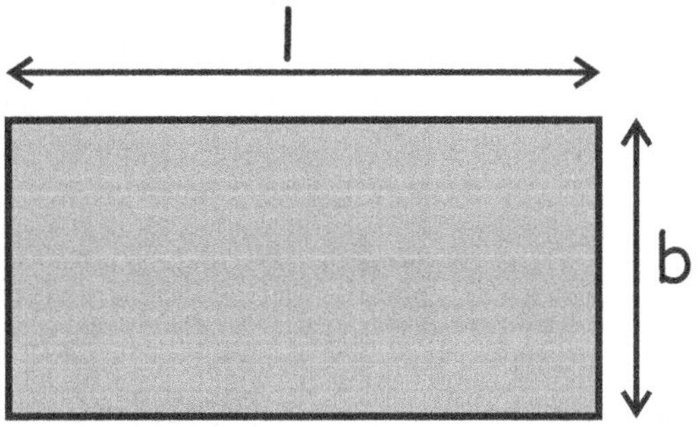

Area = l b ×

Perimeter = 2(l + b)

8. Acute angle

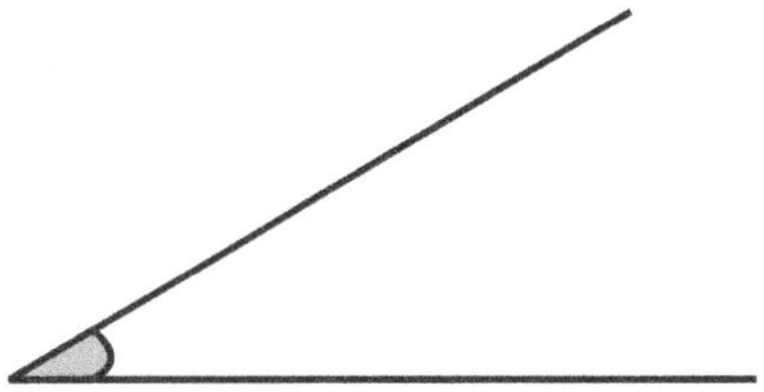

Any angle less than 90°
is called as acute angled

9. Right angle

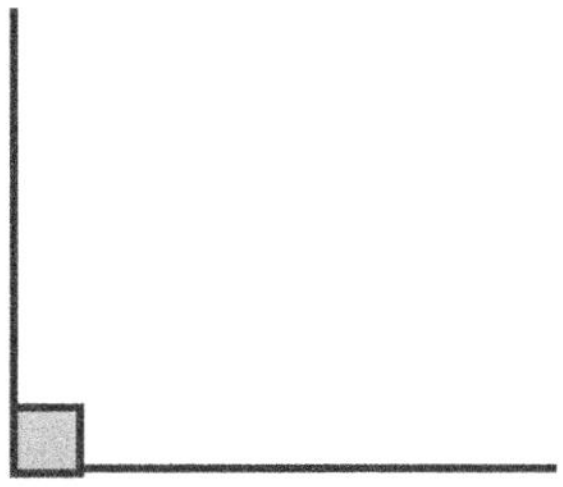

An angle equal to 90°
is called as right angled

10. Obtuse angle

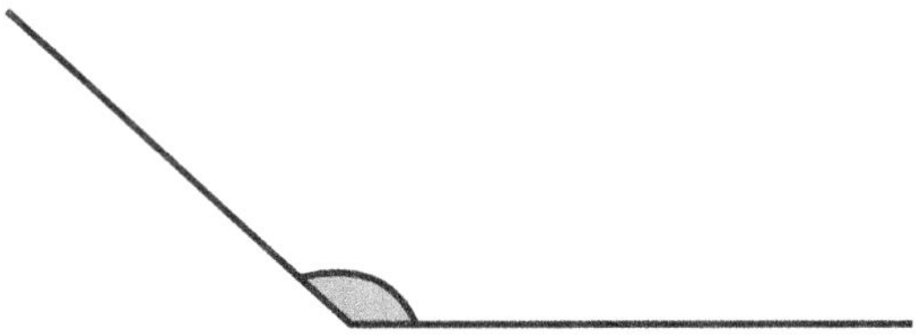

Any angle greater than 90°
is called as obtuse angled

11. Perimeter of a polygon KLMNOP

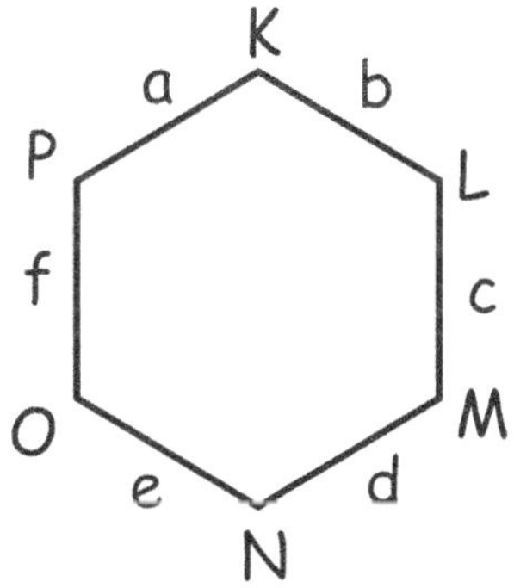

Perimeter =
(a + b + c + d + e + f) units

Abbreviations

milligram	mg
gram	g
kilogram	kg
milliliter	mL
liter	L
kiloliter	kL
millimeter	mm
centimeter	cm
meter	m
kilometer	km
square centimeter	cm^2
cubic centimeter	cm^3

year	yr
month	mon
hour	hr
minute	min
second	sec

volume	V
total Square Area	S.A
area of base	B
ounce	oz
pound	lb
quart	qt
gallon	gal.
inches	in.
foot	ft
yard	yd
mile	mi.
square inch	sq in.
square foot	sq ft
cubic inch	cu in.
cubic foot	cu ft

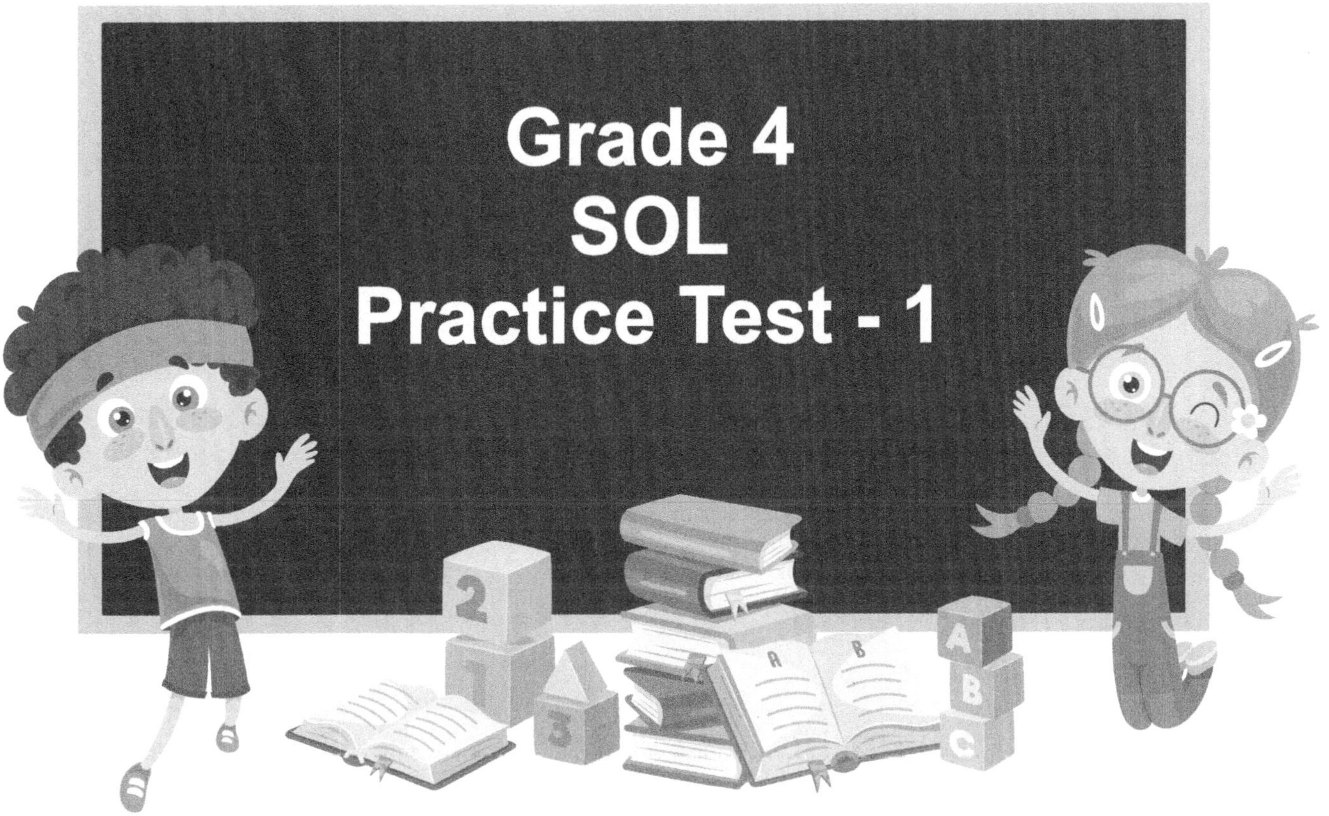
Grade 4
SOL
Practice Test - 1

1. Simplify

$$\begin{array}{r} 913,471 \\ -\ 682,190 \\ \hline \end{array}$$

(A) 231,281 (B) 231,028

(C) 231,218 (D) 231,182

2. Olivia completes reading $\frac{2}{5}$ of her 100 page adventure story. How many more pages are left to read ?

(A) 60 (B) 62

(C) 63 (D) 59

3. Cathy needs 95 beads to make a necklace. To make 3 of the same necklaces, How many beads does she need in all ?

(A) 295 (B) 285

(C) 275 (D) 35

4. Simplify $\frac{7}{9} - \frac{1}{9} = \boxed{?}$

(A) $\frac{2}{3}$ (B) $\frac{5}{9}$

(C) $\frac{4}{9}$ (D) $\frac{1}{3}$

5. Estimate $532 \div 3 = \boxed{?}$

(A) 170 (B) 174

(C) 176 (D) 177

6. Evaluate 0.99 + 0.9 = ?

(A) 11.1
(B) 1.01
(C) 1.89
(D) 11.4

7. Evaluate 1007 ÷ 53 = ?

(A) 119
(B) 109
(C) 17
(D) 19

8. Find the difference between 73564 and 40757 is ?

(A) 32800
(B) 32808
(C) 32806
(D) 32807

9. Find the product of 181 × 69 = ?

(A) 12499
(B) 12489
(C) 12480
(D) 12488

10. Evaluate $\frac{2}{5} + \frac{4}{9} =$?

(A) $\frac{38}{45}$
(B) $\frac{41}{45}$
(C) $\frac{39}{45}$
(D) $\frac{37}{45}$

11. Evaluate

$$\begin{array}{r} 82.56 \\ -\ 13.88 \\ \hline \end{array}$$

(A) 68.66 (B) 68.88

(C) 68.69 (D) 68.68

12. The difference between 717 and 319 is little less than

(A) 390 (B) 400

(C) 500 (D) 395

13. What number makes this number sentence true ?

21 + (23 + 25) = (21 + __) + 25

(A) 21 (B) 48

(C) 48 (D) 23

14. Find the product of 22 and 33. The product value is closest to ________

(A) 1726 (B) 725

(C) 726 (D) 741

15. The difference between 89,412 and 51,398 is best described as closest to _______.

(A) 38,016 (B) 38,014

(C) 38,019 (D) 39, 016

16. Find the missing number in the below pattern ?

51, 56, 61, 66,71, ?, 81

(A) 77
(B) 78
(C) 76
(D) 79

17. Choose the correct statement from the below options

(A) $11 \times 9 = 2 \times 14$
(B) $3 \times 9 = 4 \times 7$
(C) $6 \times 6 = 2 \times 10$
(D) $9 \times 9 = 27 \times 3$

18. Beth is stacking books in a pattern as shown below. She piles six stacks of books. How many books are there in all the six stacks?

(A) 21
(B) 22
(C) 25
(D) 20

19. Fill the series by finding the missing number ?

507, 500, 493, 486, ?

(A) 488
(B) 478
(C) 479
(D) 480

20. Which of the below number sentence is not true ?

(A) $7 + (3 + 2) = (1 + 3) + 7$

(B) $\frac{54}{3} = 18 \times 1$

(C) $22 + 2 = 10 + 14$

(D) $(6 \times 1) \times 7 = 6 \times (1 \times 7)$

21. Which number sentence is true ?

(A) $13 + (11 + 6) = 6 + (13 + 11)$

(B) $1 + (2 + 3) = 3 + (2 - 1)$

(C) $11 - (3 + 5) = 5 + (11 + 3)$

(D) $20 + (6 - 3) = 3 + (20 - 6)$

22. Complete the number sentence below

$$111 + (222 + 333) = (222 + 111) + ?$$

(A) 111

(B) 212

(C) 333

(D) 222

23. Complete the pattern

1150, 2250, 3350, ______, _______

(A) 4450, 5550

(B) 4440, 5550

(C) 4450, 5540

(D) 4450, 5450

24. Below table shows the donuts making time by a ding dong machine. If the pattern shown in the table continues in the same way, how much time will it take to make three dozens of donuts ?

Number of Donuts	Time in Minutes
1	2
2	4
3	6
4	8

(A) 80 (B) 72

(C) 76 (D) 75

25. A box contains 13 pencils,21 erasers, 11 crayons and 7 pens and 1 book marker. Bella picks up an item from the box. Which of the below describes the chances of picking a book marker from the box ?

(A) Unlikely, but not impossible (B) Impossible

(C) Likely, but not certain (D) Certain

26. A box contains 7 red cars, 13 white cars and 15 green cars. Liam picks up a car from the box. What is the probability of picking a red car ?

(A) $\frac{3}{7}$ (B) $\frac{13}{35}$

(C) $\frac{1}{5}$ (D) $\frac{7}{28}$

27. What is the maximum number of line segments shown in the figure below ?

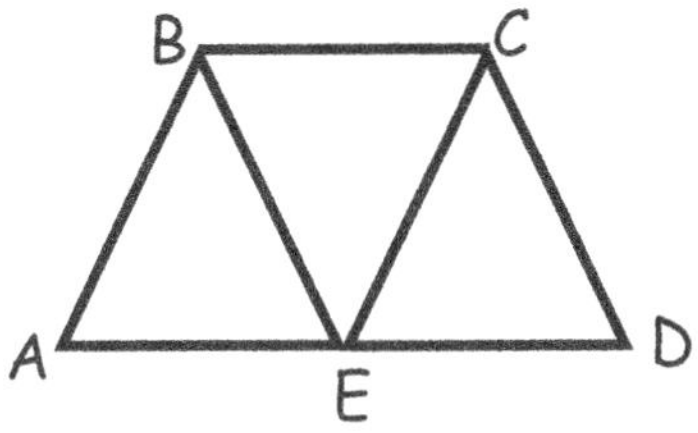

(A) 10
(B) 11
(C) 7
(D) 8

28. Which of the below is the correct statement for the Parallelogram KLMN ?

(A) Four right angles
(B) Opposite sides are parallel
(C) Opposite sides are not congruent
(D) All sides are equal

29. Estimate the weight of the toy cars

(A) 8 lbs
(B) 5 lbs
(C) 10 lbs
(D) 4 lbs

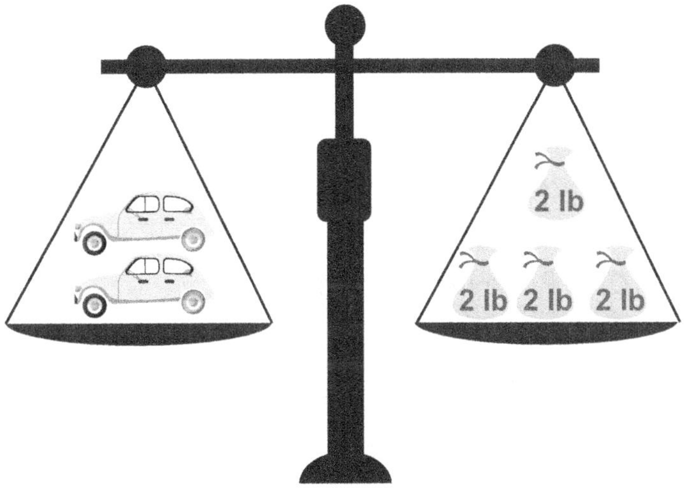

30. Height of the ladder is 3 meters. What is its approximate length in feet ?

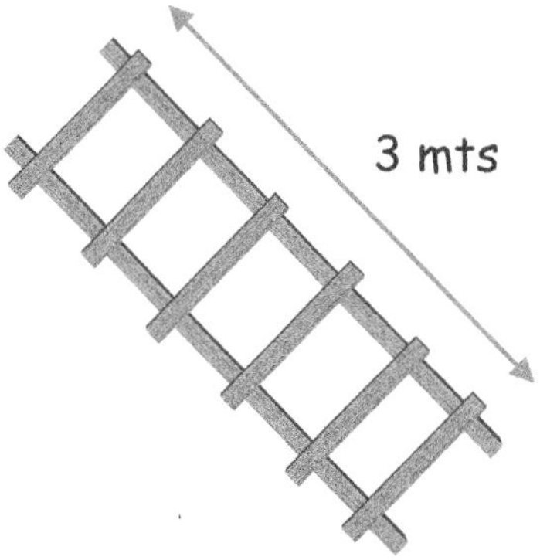

(A) 9.84 feet

(B) 9.98 feet

(C) 10 feet

(D) 8.84 feet

31. Which pair of the below lines are intersecting ?

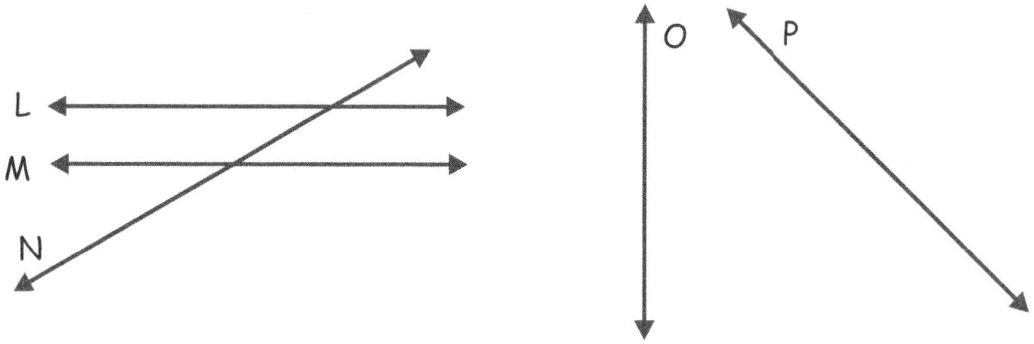

(A) O, P

(B) M, O are intersecting

(C) L, P are intersecting

(D) L, N are intersecting

32. Caterpillar weighs 27 grams. What is the caterpillar's weight in milligrams ?

(A) 270 milligrams

(B) 27000 milligrams

(C) 2700 milligrams

(D) 270000 milligrams

33. The below model represents the fraction $\frac{8}{12}$

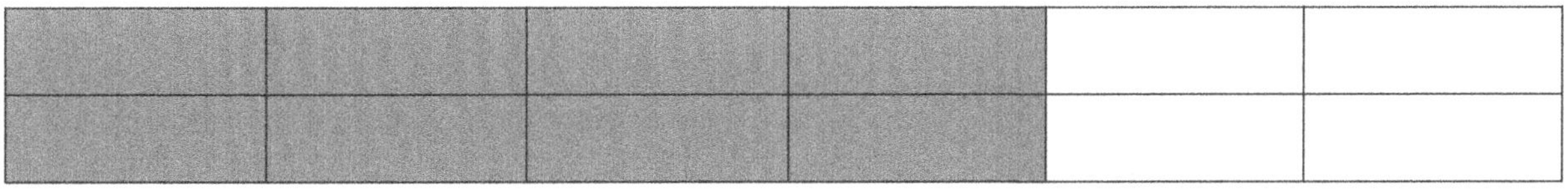

Which of the below is also equal to $\frac{8}{12}$?

(A)

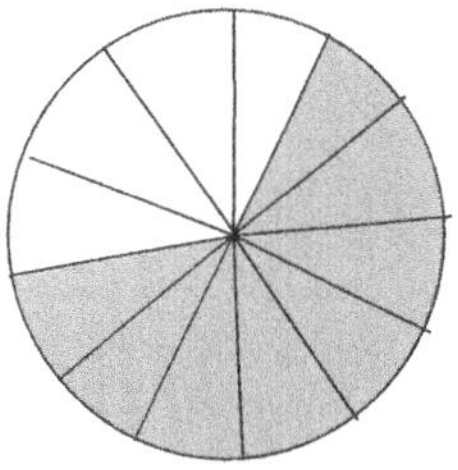

(B)

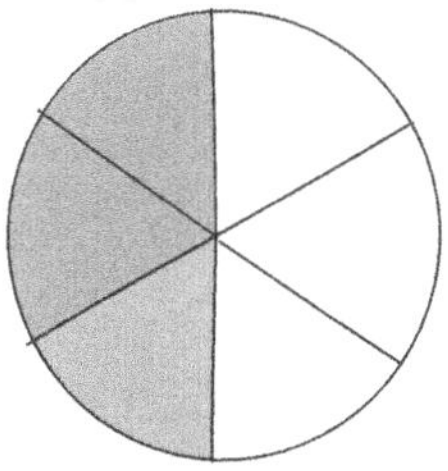

(C)

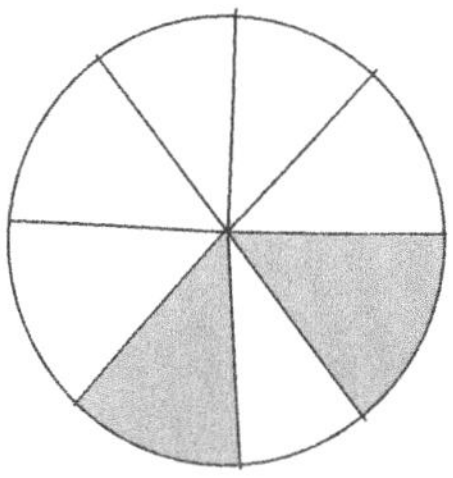

(D)

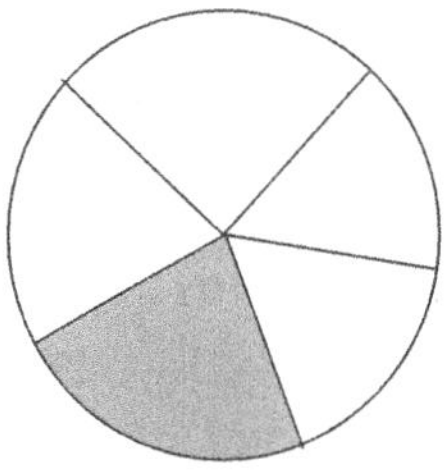

34. If below model equals to 1

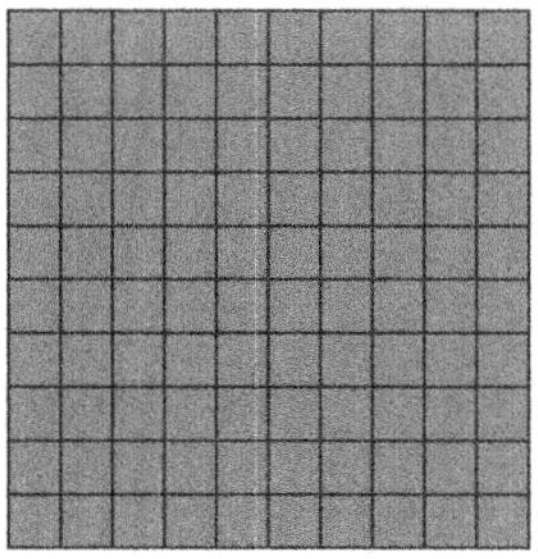

then

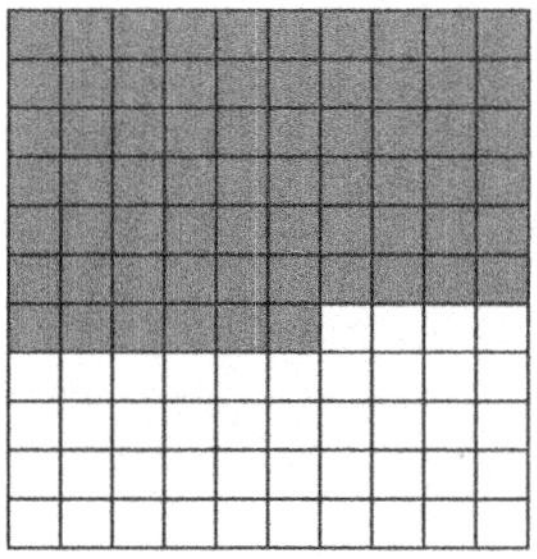

(A) 0.066 (B) 0.065

(C) 0.66 (D) 0.0066

35. Which of the below will round to 358.21 ?

(A) 358.2059 (B) 358.2001

(C) 357.201 (D) 359.2100

36. Which of the below fractions has least value ?

(A) $\frac{3}{4}$ (B) $\frac{4}{6}$

(C) $\frac{3}{7}$ (D) $\frac{5}{9}$

37. Round 97,137,232 to the nearest million.

(A) 97,140,000 (B) 97,137,200

(C) 98,000,232 (D) 97,000,000

38. Round 99,093 to the nearest ten.

(A) 99,093 (B) 99,100

(C) 99,010 (D) 99,090

39. Which of the below statement is correct ?

(A) $7706 < 7806$ (B) $7709 < 7008$

(C) $4309 = 4390$ (D) $7001 > 7077$

40. Write 606,623 in words.

(A) Six hundred sixteen thousand, six hundred twenty-three
(B) Six hundred sixty-six thousand, six hundred twenty-three
(C) Sixty-seven thousand, six hundred and six
(D) Six hundred and six thousand, six hundred twenty-three

41. Which of the below statements is correct ?

(A) $\frac{5}{7} = \frac{10}{11}$

(B) $\frac{5}{6} > \frac{2}{5}$

(C) $\frac{3}{9} = \frac{4}{11}$

(D) $\frac{5}{7} < \frac{2}{5}$

42. Which ordered pair would have represented the square/parallelogram ?

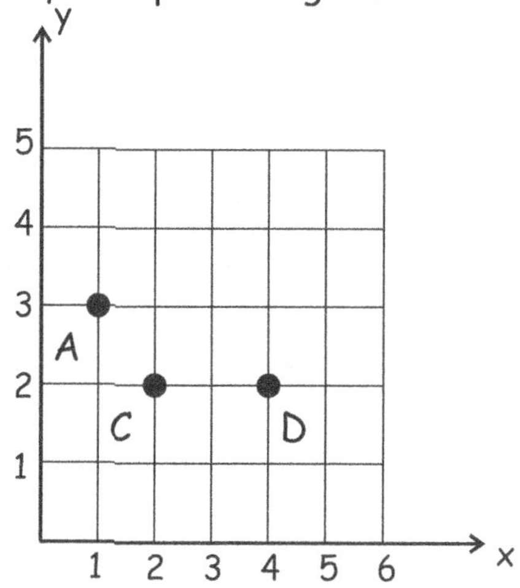

(A) (3,3)

(B) (2,3)

(C) (4,4)

(D) (3,4)

43. Dan measures apples on the balance scale. Based on the information given below estimate the weight of one apple.

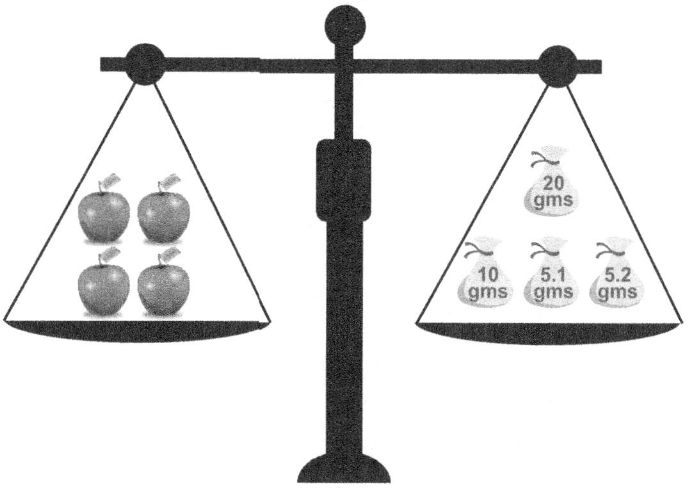

(A) 5 gms

(B) 10 gms

(C) 5.3 gms

(D) 20.15 gms

44. Estimate the perimeter of the triangle given below ?

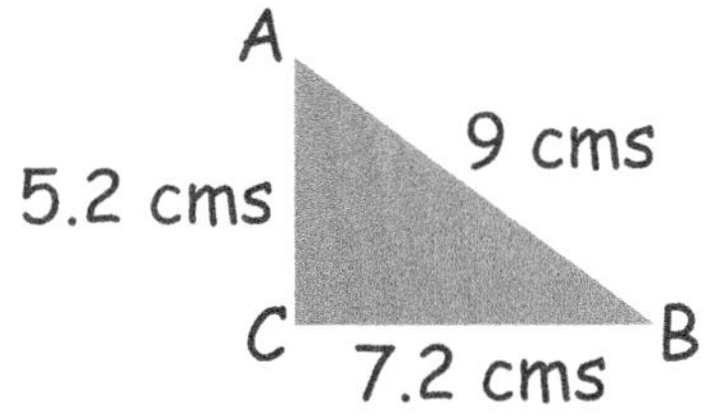

(A) 22 cms (B) 17.5 cms

(C) 21 cms (D) 35 cms

Graph problems (Histogram)

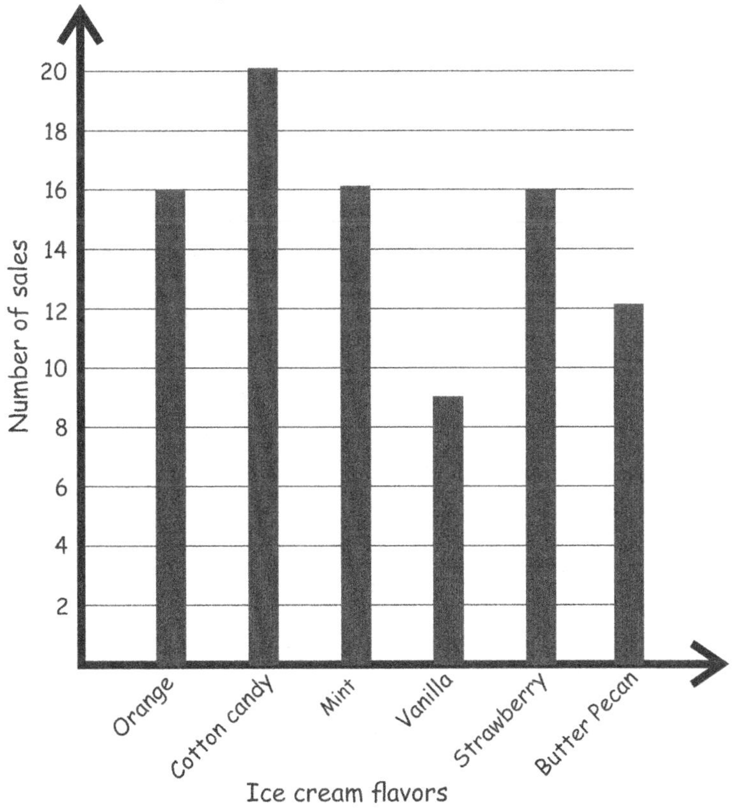

45. Which Ice cream flavor had highest number of sales ?

(A) Strawberry (B) Orange
(C) Cotton candy (D) Butter Pecan

46. Which Ice cream flavor had least number of sales?

(A) Butter Pecan
(B) Mint
(C) Cotton candy
(D) Vanilla

47. How many Ice creams were sold ?

(A) 89
(B) 77
(C) 65
(D) 91

48. Estimate the length of this book ?

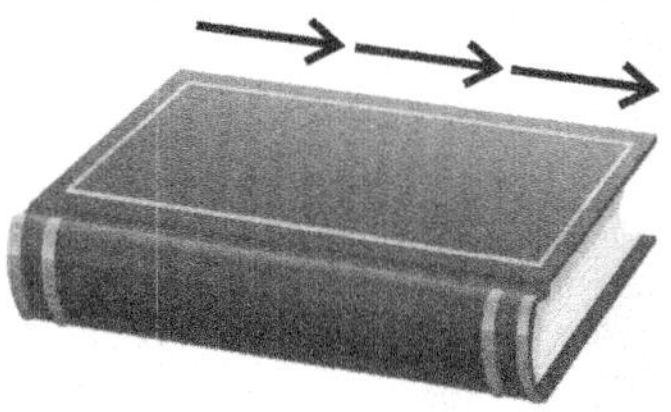

KEY :
→ = 1.5 cms

(A) 4 cms
(B) 4.5 cms
(C) 5 cms
(D) 6 cms

49. Which of the below pair are congruent ?

(A) 

(B) 

(C) 

(D) 

50. Which of the below pairs of figures are congruent ?

(A) 

(B) 

(C) 

(D) 

Grade 4
SOL
Practice Test - 2

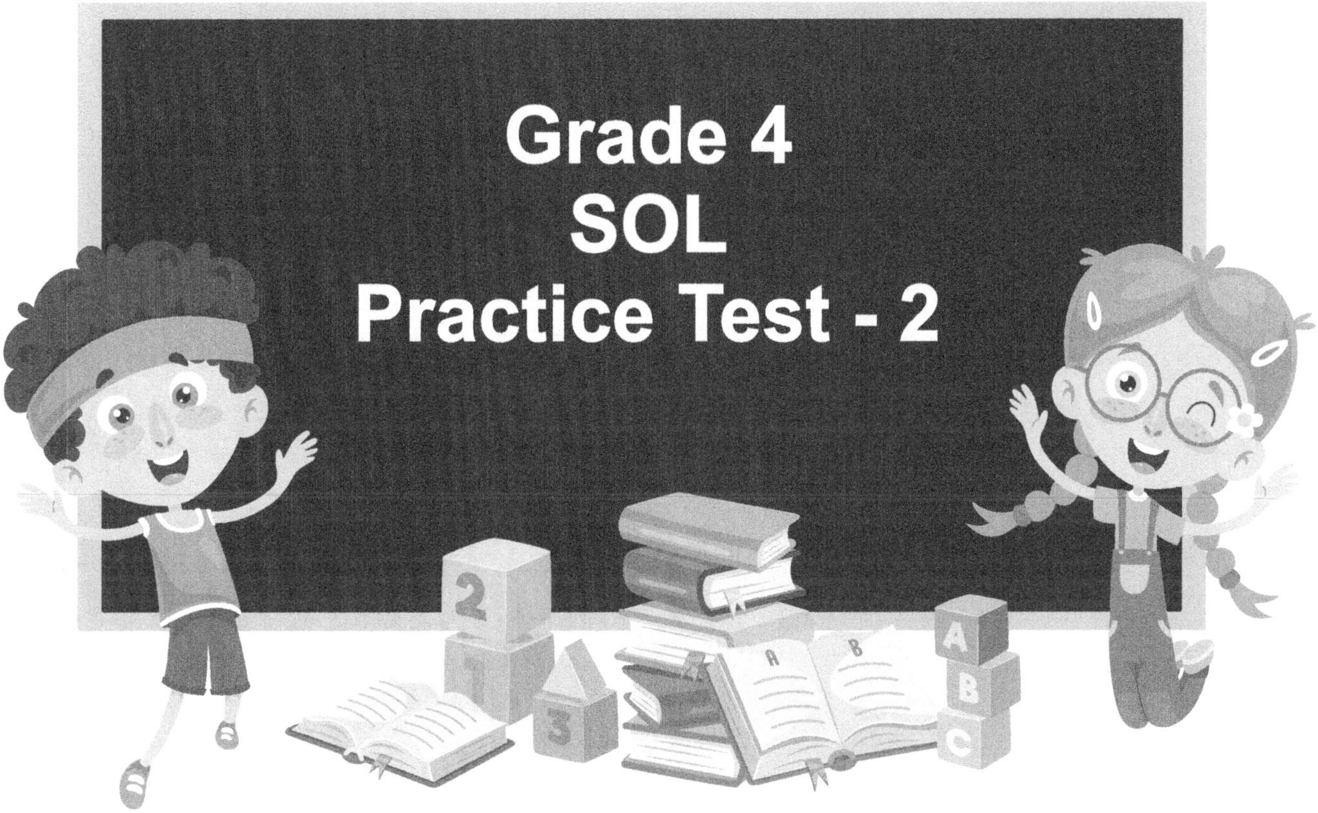

1. Simplify

$$\begin{array}{r} 925,945 \\ -\ 234,829 \\ \hline \end{array}$$

(A) 697,126

(B) 691,116

(C) 691,126

(D) 619,119

2. Which of the below product is between 950 and 1000 ?

(A) 37 × 35

(B) 39 × 31

(C) 43 × 11

(D) 31 × 32

3. Simplify $\frac{5}{13} + \frac{2}{3} = \boxed{?}$

(A) $\frac{41}{39}$

(B) $\frac{15}{397}$

(C) $\frac{41}{13}$

(D) $\frac{26}{39}$

4. Cathy buys a sand witch for $2.35, chips for $0.30 and soda for $1.09.
She gives the cashier a $5. How much will she get back from the cashier ?

(A) 1.32

(B) 1.27

(C) 1.26

(D) 1.31

5. Below model is equal to 1.

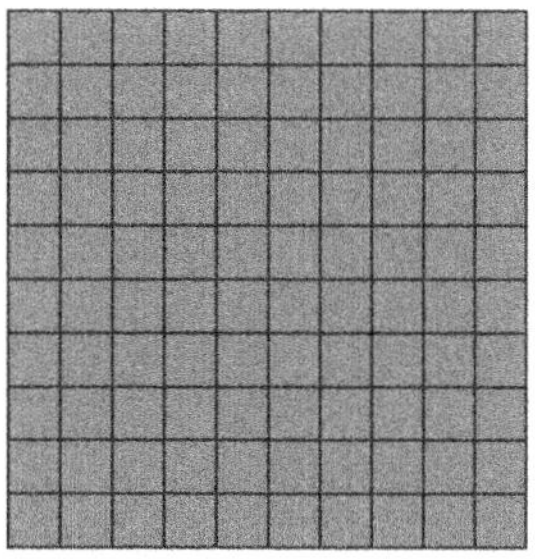

Which decimal number is represented by the below model.

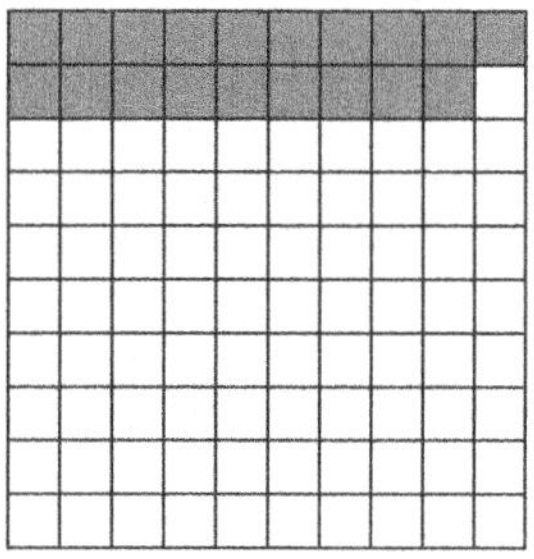

(A) 0.17 (B) 0.18

(C) 0.19 (D) 0.13

6. Evaluate 37.41 - 19.09

(A) 18.31 (B) 18.32

(C) 18.35 (D) 18.33

7. Simplify $\frac{1201}{11} = \boxed{?}$

(A) 1099 R 3 (B) 109 R 6

(C) 109 R 4 (D) 109 R 2

8. Which best describes the difference 2,927 - 1001 ?

(A) Closer to 2000 than 2500

(B) Closer to 3000 than 3500

(C) Closer to 1000 than 2000

(D) Closer to 3500 than 4000

9. Evaluate $\frac{995}{3} = \boxed{?}$

(A) 331 R 0

(B) 331 R 1

(C) 331 R 2

(D) 332 R 2

10. Find the product of 89 and 27

(A) 2403

(B) 2400

(C) 2402

(D) 2405

11. Which of the below pair are congruent to each other ?

(A) 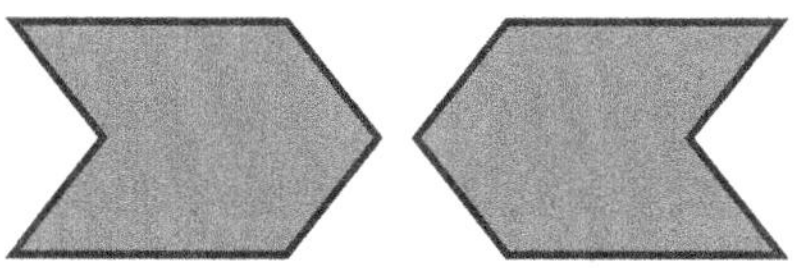

(B) 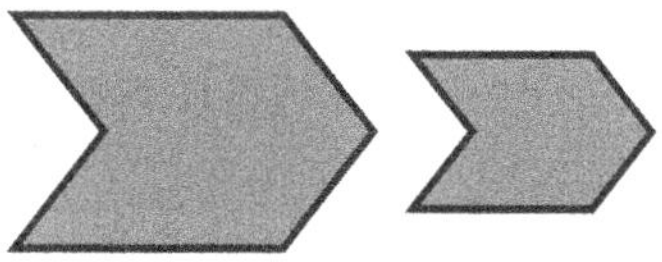

(C) 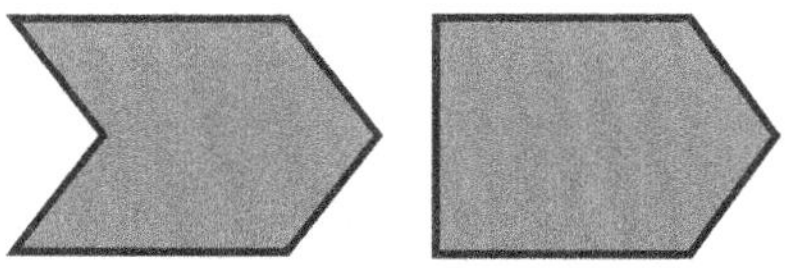

(D) 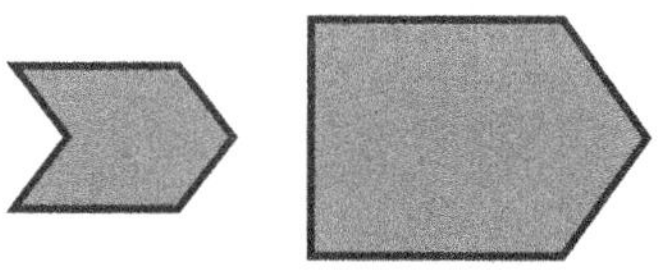

12. Evaluate $\frac{5559}{11} = \boxed{?}$

(A) 505 R 3

(B) 55 R 4

(C) 5005 R 3

(D) 505 R 4

13. Which two factors have a product of 936 ?

(A) 24 × 34

(B) 36 × 26

(C) 56 × 26

(D) 34 × 24

14. Round 190,901 to the nearest hundred.

(A) 190,900

(B) 190,000

(C) 191,000

(D) 199,900

15. Write "seventeen and five thousandths" in standard form.

(A) 17.05

(B) 170.005

(C) 17.005

(D) 107.05

16. Which of the following shows a fraction model and a decimal model that both represent the same value ?

(A) 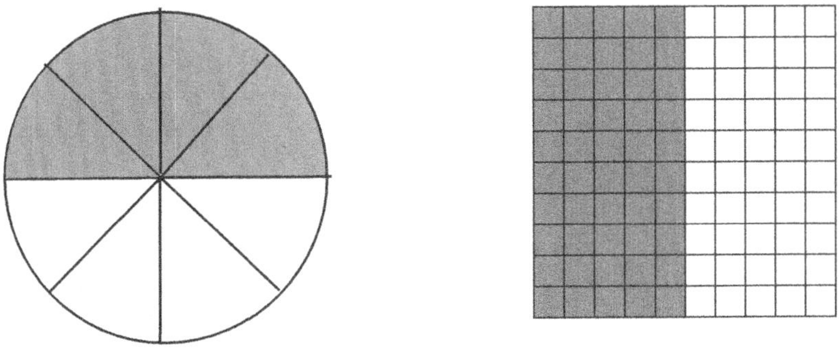

(B) 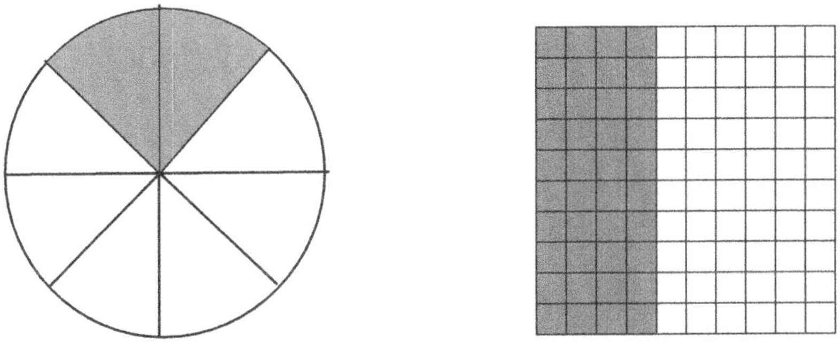

(C) 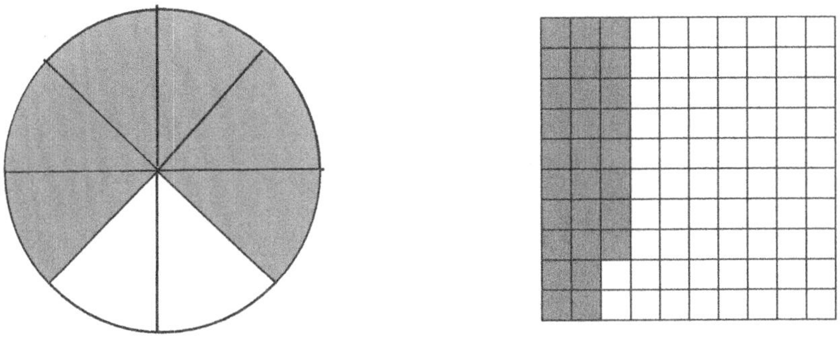

(D) 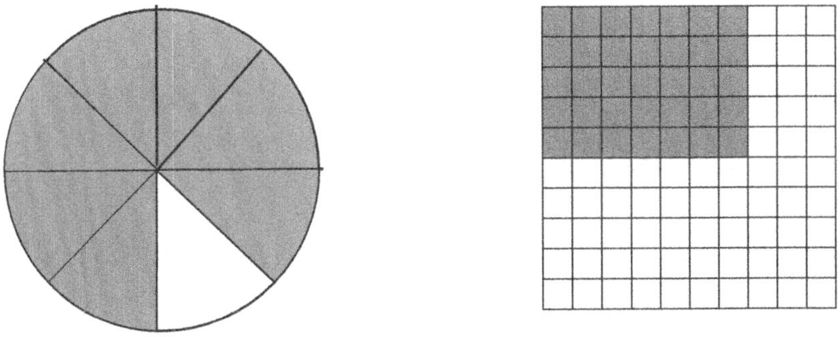

17. Which of the below model is greater than $\frac{5}{11}$?

(A) 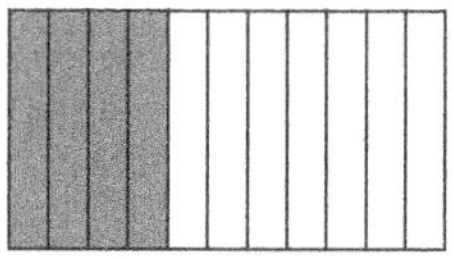

(B) 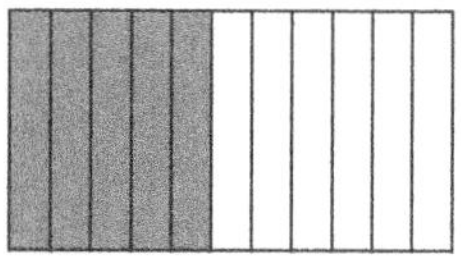

(C) 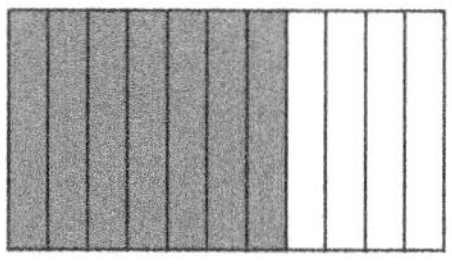

(D) 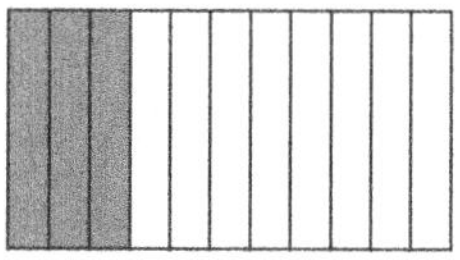

18. Which of the below is a true statement ?

(A) 719065 < 719005

(B) 719065 > 719105

(C) 719165 = 719265

(D) 719065 > 719005

19. Point X is located at $\frac{5}{7}$ on the number line

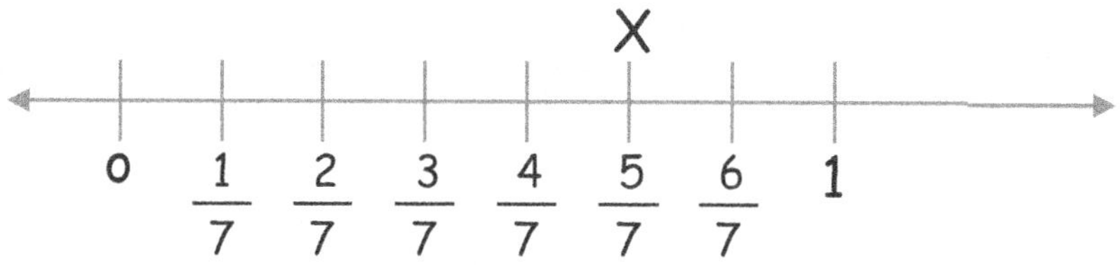

Which of the following is another name for location of X?

(A) $\frac{2}{3}$

(B) $\frac{10}{14}$

(C) $\frac{2}{5}$

(D) $\frac{11}{14}$

20. Which of the below statement is true ?

(A) 1.094 = 1.0094

(B) 1.093 < 0.093

(C) 1.093 > 1.094

(D) 1.093 > 1.091

21. The shaded sections of these models represent two decimal numbers.

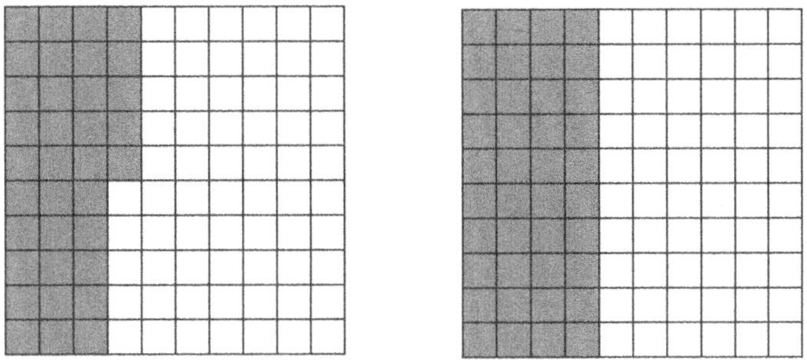

Based on the above which of the below is true ?

(A) 0.35 < 0.40

(B) 0.36 = 0.40

(C) 0.36 < 0.30

(D) 0.35 > 0.40

22. Which of the below shows line segment ?

(A)

(B) 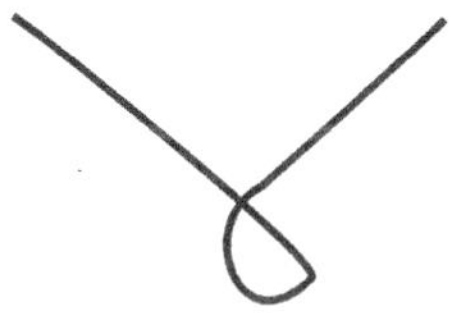

(C) 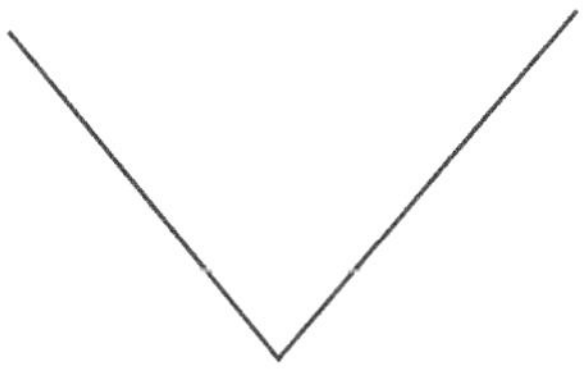

(D) 

23. Which of the following is the closest to the height of the building shown ?

(A) 2.5 ft

(B) 3.0 ft

(c) 4.5 ft

(D) 6.0 ft

24. Rosy is making a 2 feet long table. How long is the new table in inches ?

(A) 24 inches
(B) 22 inches
(C) 20 inches
(D) 18 inches

25. Ms.Lora's fruit garden is 525 meters length wise. How long is the fruit garden in centimeters ?

(A) 525 cm
(B) 52500 cm
(C) 525000 cm
(D) 5250 cm

26. Which of the below is a true statement for a line ?

(A) It has one end point and extends on other way
(B) It has no end points
(C) It has a starting point
(D) It has two end points

27. Which of the below pair are perpendicular lines ?

(A) 

(B) 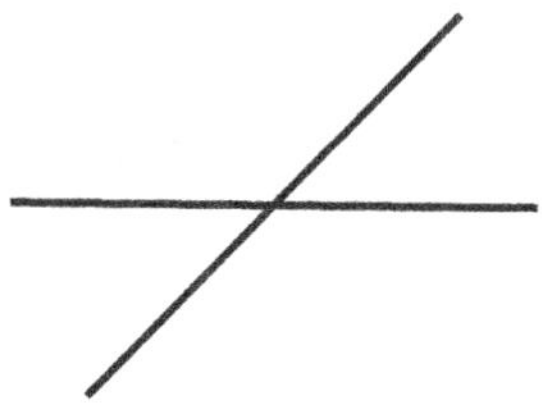

(C) 

(D) 

28. Sony is making cupcakes which requires 3 pints of butter. How much butter does she need in cups ?

(A) 4 cups

(B) 3 cups

(C) 6 cups

(D) 9 cups

29. Ella the civil engineer built a beautiful building of height 1320 inches. How high is the building in feet ?

(A) $\frac{1320}{12}$

(B) 1320 × 10

(C) 1320 × 12

(D) $\frac{1320}{10}$

30. Thomas got 2.5 pounds of almond milk. Which of the below is also equivalent to 2.5 pounds of almond milk ?

(A) 100 grams

(B) 250 ounces

(C) 40 ounces

(D) 38 ounces

31. Which order pair represents the missing coordinate of the below square ABCD ?

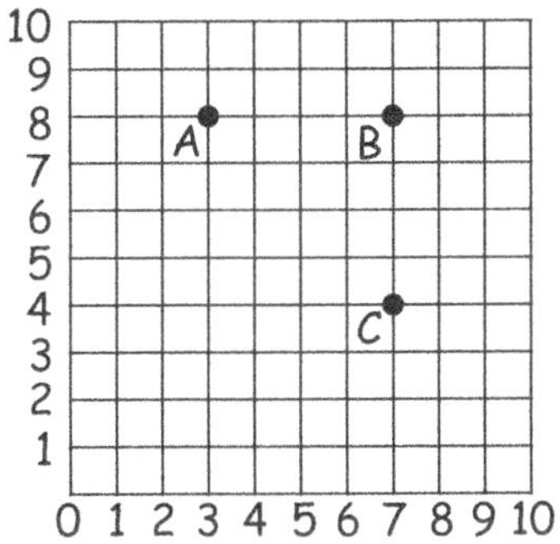

(A) (3,3)

(B) (3,4)

(C) (4,3)

(D) (4,4)

32. Below graph shows the number of votes received from Hickary elementary school on a survey conducted.

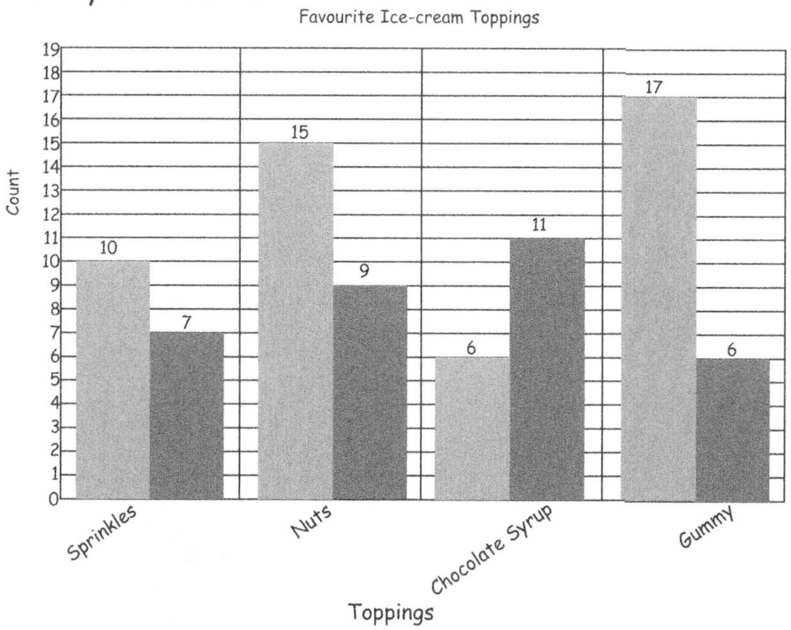

Based on the data in the graph, which of the following is closest to the total number of kids who liked Chocolate Syrup as the toppings ?

(A) 20 (B) 19

(C) 24 (D) 17

33. Eliza has a candy bag with the below candies:

Skittle	7 packs
Gummy Bear	13 packs
Sour Patch	21 packs
M & M's	1 pack

Max picks one candy pack from the bag without looking in. Which candy is least likely to be picked up ?

(A) Skittle (B) Sour Patch

(C) M & M's (D) Gummy Bear

34. A bag contains 2 red, 3 green and 2 blue balls. What is the probability that none of the balls drawn is blue ?

(A) $\frac{7}{21}$ (B) $\frac{11}{21}$

(C) $\frac{2}{7}$ (D) $\frac{5}{7}$

35. The table below shows various fruits that are distributed to senior citizens in the

Apples	121
Oranges	95
Bananas	58
Pears	77

Daisy receives one of the above. What is the chance of her getting an apple ?

(A) Unlikely, but not impossible (B) Impossible

(C) Likely, but not certain (D) Certain

36. Which of the below figures are congruent to each other ?

(A) 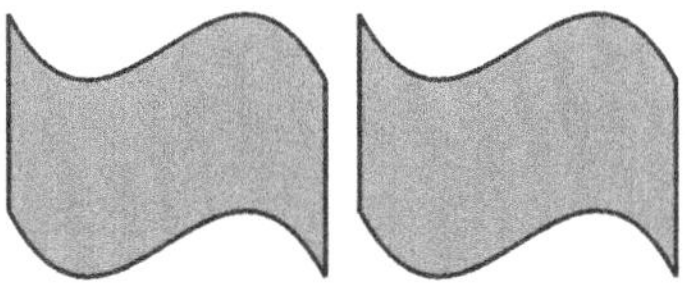

(B) 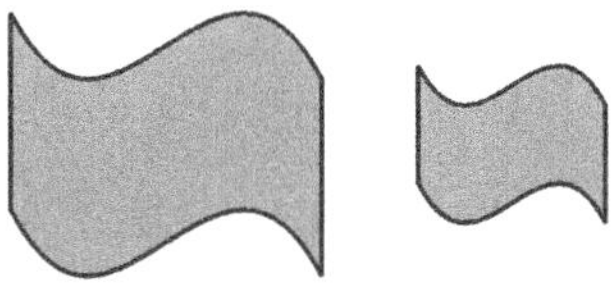

(C) 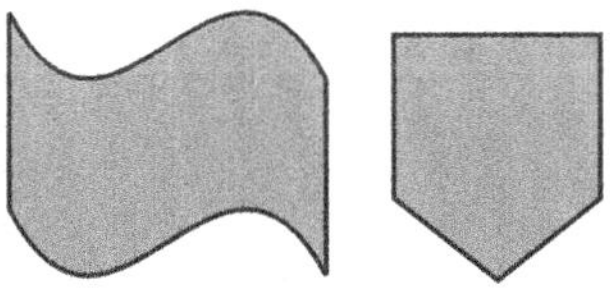

(D) 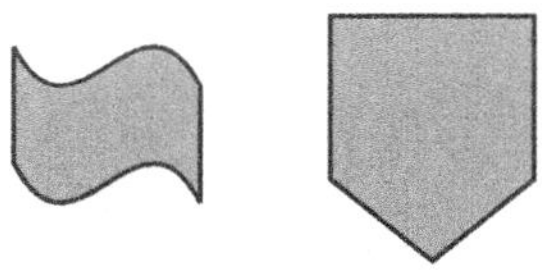

37. The bar graph shows the number of students that have visited the spy museum over a two and a half-hour period.

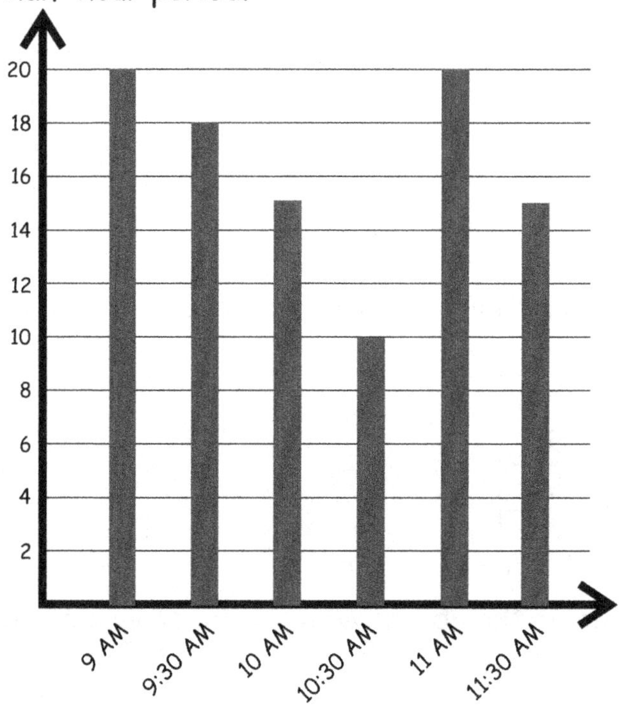

According to the graph. How many students have visited the Science museum at before 10 AM.

(A) 38

(B) 41

(C) 36

(D) 47

38. Which of the following is an example of reflection ?

(A) 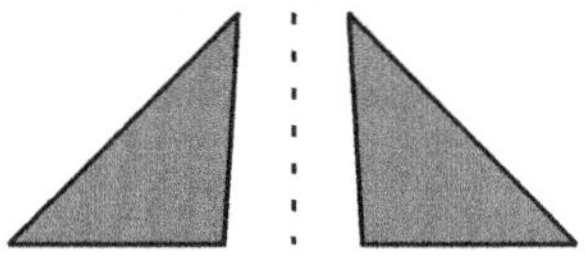

(B) 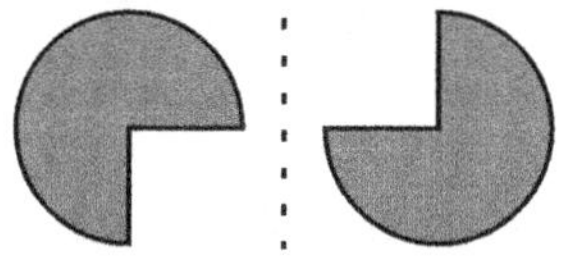

(C) 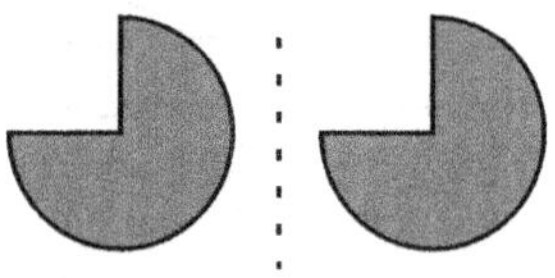

(D) 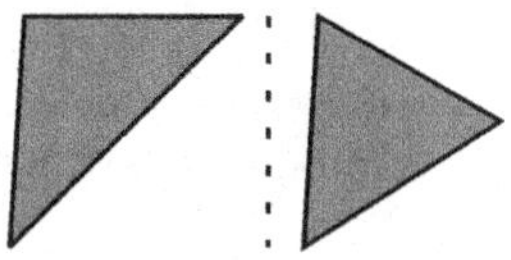

39. Iris will pick one tile from a bag without looking. From which of the following bags is she certain to pick a tile with a "11" on it ?

(A)

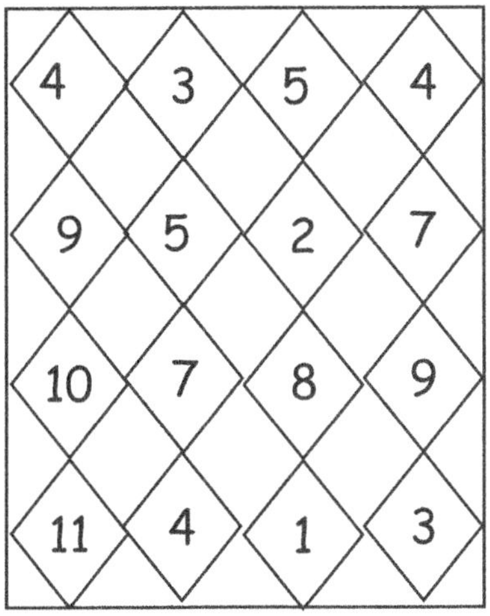

(B)

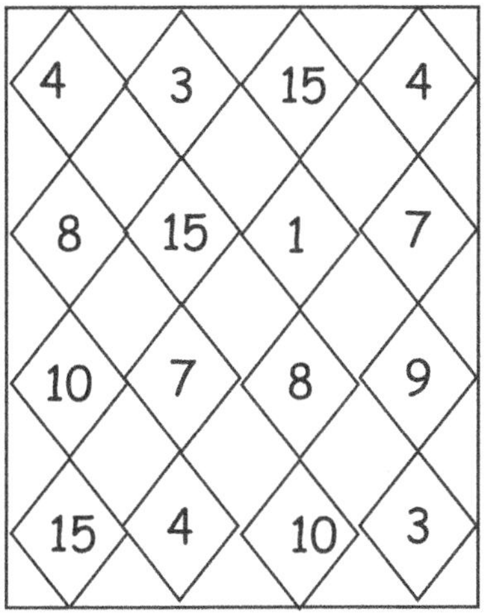

(C)

(D)

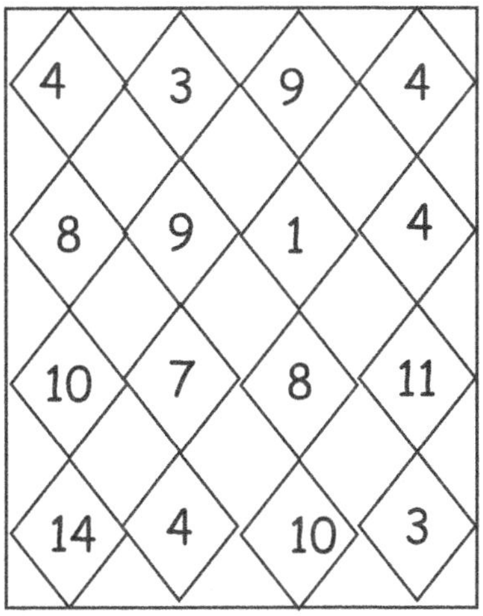

40. Oscar is tossing two coins. What is the probability that he gets a head on both coins ? (order doesn't matter)

(A) $\frac{1}{4}$

(B) $\frac{2}{5}$

(C) $\frac{1}{3}$

(D) $\frac{5}{6}$

41. Which of the below is correct ?

(A) 14 + (16 - 15) = (14 + 16) + 15

(B) (14 - 15) + 16 = 14 + (15 - 16)

(C) 14 × (15 - 16) = (14 × 15) - 16

(D) 14 + (15 + 16) = (14 + 15) + 16

42. Mrs. G ordered snack boxes in bulk that contain various varieties of snacks. The table below shows the total number of boxes and the number of snack packs it contains

Box #	Number of snack packs
1	128
2	112
3	252
4	448

Based on the given data in the table, what will be the total number of snacks in all boxes ?

(A) 924

(B) 940

(C) 944

(D) 280

43. Which of the below statements is correct ?

(A) $2 \times 9 = 13 \times 3$

(B) $2 \times 9 = 9 + 2$

(C) $2 \times 9 = 11 + 7$

(D) $13 \times 9 = 19 \times 13$

44. Line segment PQ is marked using the rule "add 5" to mark points in a pattern on the line. Based on the rule what will be point A marked as ?

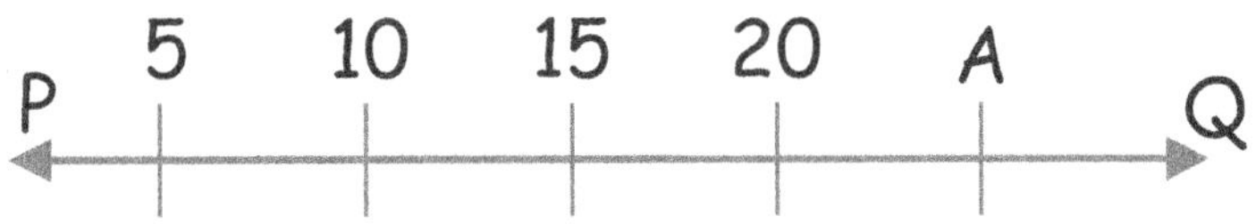

(A) 25

(B) 30

(C) 20

(D) 35

45. Which of the below is correct ?

(A) $(122 \times 207) - 201 = 125 - (207 \times 201)$

(B) $\dfrac{(122 \times 207)}{201} = \dfrac{122}{(207 \times 201)}$

(C) $(122 \times 207) + 201 = 122 + (207 \times 201)$

(D) $(122 \times 207) \times 201 = 122 \times (207 \times 201)$

46. The points on this number line represent a pattern.

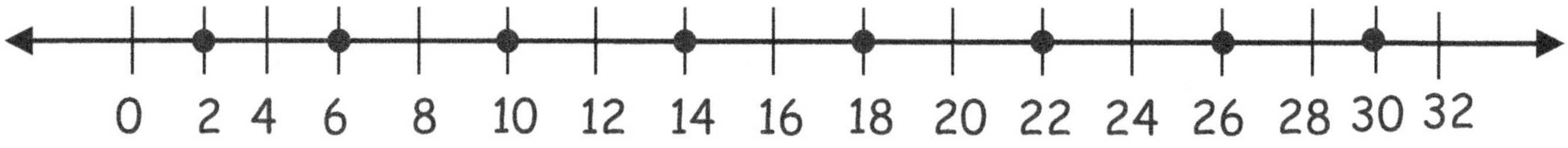

Which of these number lines continues this pattern of points in the same way ?

(A)

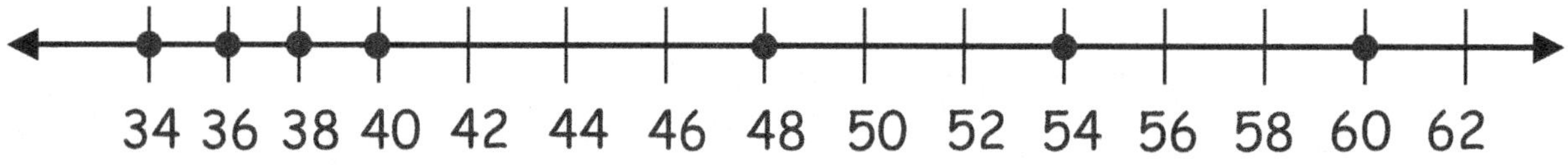

(B)

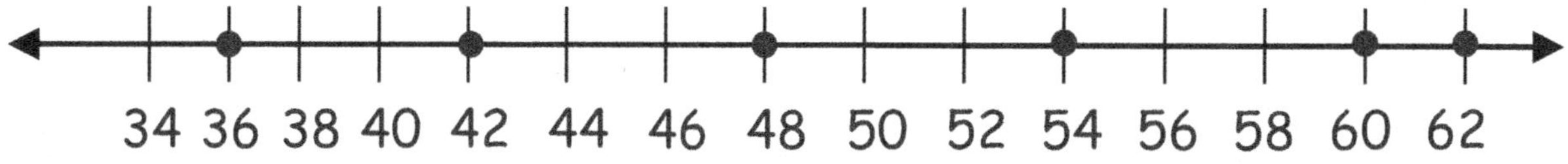

(C)

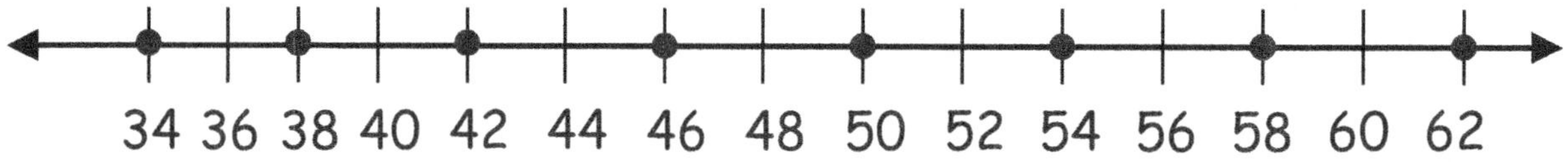

(D)

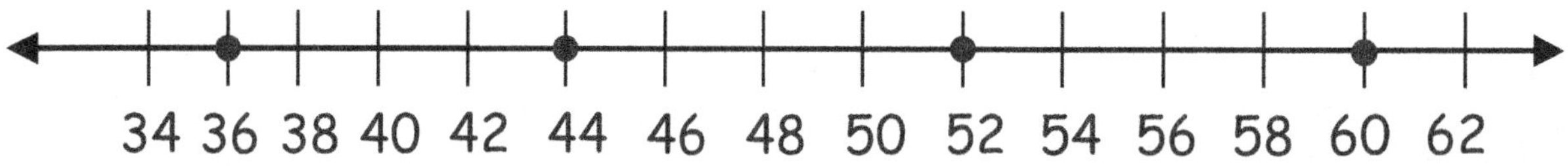

47. Find the next three numbers in the below pattern.

42 , 53 , 64 , 75 , ...

(A) 83 , 94 , 105

(B) 84 , 95 , 106

(C) 86 , 96 , 108

(D) 86 , 97 , 108

48. Based on the below table find the rule used on In

In	Out
45	15
30	10
15	5

(A) Multiply by 2

(B) Divide by 3

(C) Add 4

(D) Add 5

49. Which of the below is correct ?

(A) $4 \times 5 = 2 + 18$

(B) $4 \times 5 = 18 - 2$

(C) $4 \times 5 = \frac{18}{2}$

(D) $4 \times 5 = 4 + 5$

50. Which of the below is correct ?

(A) $\frac{91}{13} = 91 + 13$

(B) $\frac{91}{13} = 4 - 3$

(C) $\frac{91}{13} = \frac{10}{2}$

(D) $\frac{91}{13} = 4 + 3$

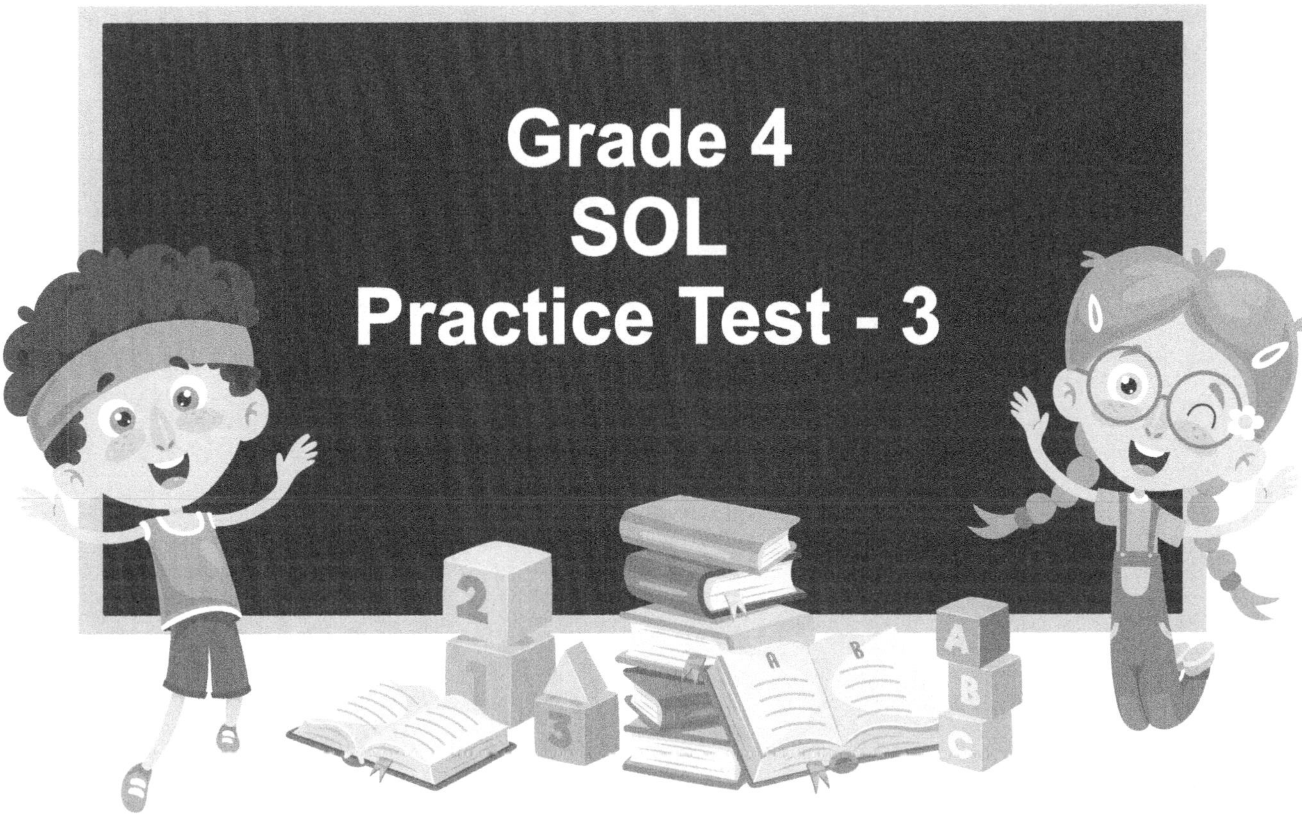
Grade 4
SOL
Practice Test - 3

1. Subtract

$$\begin{array}{r} 76,367 \\ -\ 32,354 \\ \hline \end{array}$$

(A) 44,030 (B) 44,031

(C) 44,013 (D) 44,103

2. Super dog gives birth to a puppy which weighed $\frac{5}{6}$ kilograms. After two weeks, its weight increased by $\frac{1}{10}$ kilogram. Puppy got sick and lost weight of $\frac{1}{5}$ kilograms. How much does the puppy weigh now ?

(A) $\frac{20}{39}$ (B) $\frac{22}{60}$

(C) $\frac{21}{30}$ (D) $\frac{22}{30}$

3. A zig saw puzzle has 101 pieces. Nora has 6 of those puzzles. How many puzzle pieces does he have in all ?

(A) 95 (B) 107

(C) 606 (D) $\frac{101}{6}$

4. Find the difference between $\frac{5}{7} - \frac{3}{7} = \boxed{?}$

(A) $\frac{5}{7}$ (B) $\frac{2}{7}$

(C) $\frac{4}{7}$ (D) $\frac{1}{7}$

5. Estimate $\frac{792}{3}$ = ?

(A) 269
(B) 265
(C) 260
(D) 264

6. Evaluate 7.28 + 5.12 = ?

(A) 12.2
(B) 12.5
(C) 11.3
(D) 12.4

7. Evaluate $\frac{144}{8}$ = ?

(A) 18
(B) 16
(C) 118
(D) 72

8. Find the difference between 14150 and 7176 is ?

(A) 26230
(B) 6974
(C) 26232
(D) 262234

9. Find the product of 994 × 22 = ?

(A) 21868
(B) 21888
(C) 21866
(D) 21858

10. Simplify the below

$$\frac{12}{35} + \frac{4}{7} = \boxed{?}$$

(A) $\frac{33}{35}$

(B) $\frac{32}{35}$

(C) $\frac{30}{35}$

(D) $\frac{34}{35}$

11. Evaluate

$$\begin{array}{r} 54.34 \\ -\ 27.71 \\ \hline \end{array}$$

(A) 26.63

(B) 26.66

(C) 26.60

(D) 26.03

12. Find the difference between 399 and 251 is little more than

(A) 150

(B) 400

(C) 100

(D) 140

13. What number makes this number sentence true ?

9 + (17 + 1) = (9 + ___) + 1

(A) 9

(B) 1

(C) 17

(D) 18

14. The product of 29 × 29 is closest to ________

(A) 844

(B) 5813

(C) 839

(D) 841

15. The difference between 227,399 and 102,659 is best described as closest to ____.

(A) 124,740
(B) 124,754
(C) 125,740
(D) 124,744

16. Find the missing number in the below pattern ?

21, 25, 29, 33, 37, ?, 45

(A) 38
(B) 41
(C) 40
(D) 42

17. Choose the correct statement from the below options

(A) 7 × 9 = 3 × 21
(B) 8 × 9 = 4 × 9
(C) 4 × 5 = 3 × 12 ...
(D) 3 × 6 = 2 × 11

18. If the below pattern continues to five figures, how many stars are used all together ?

(A) 25
(B) 29
(C) 37
(D) 35

19. Fill the series by finding the missing number ?
452, 448, 444, 440, 436, ?

(A) 431 (B) 428

(C) 430 (D) 432

20. Which of the below number sentence is not true ?

(A) 2 + (3 + 4) = (2 + 3) + 4 (B) $\frac{99}{2} = 33 \times 1$

(C) 22 - 2 = 20 - 0 (D) (7 × 3) × 1 = 7 × (3 × 1)

21. Which of the below number sentence is true ?

(A) 13 + (11 + 6) = 6 + (13 - 11) (B) 1 + (2 + 3) = 3 + (2 - 1)

(C) 11 + (3 + 5) = 5 + (11 + 3) (D) 20 + (6 - 3) = 6 + (20 + 3)

22. Complete the number sentence below
121 + (132 + 143) = (121 + 143) + ?

(A) 121 (B) 143

(C) 132 (D) 133

23. Find the figure that does not belong to the pattern from the below:

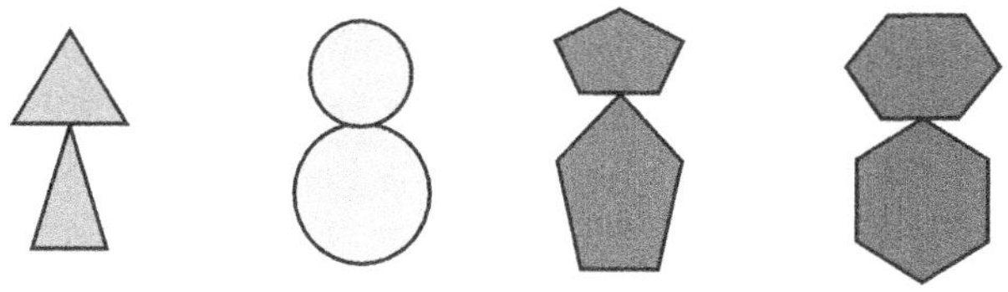

(A) 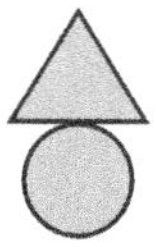

(B) 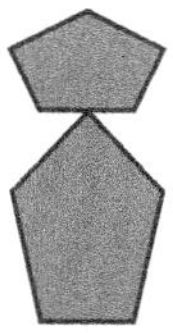

(C) 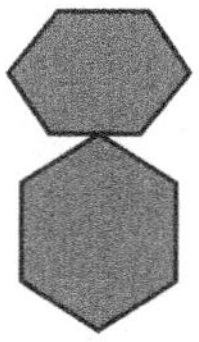

(D) 

24. Which of the below pairs of figures congruent ?

(A) 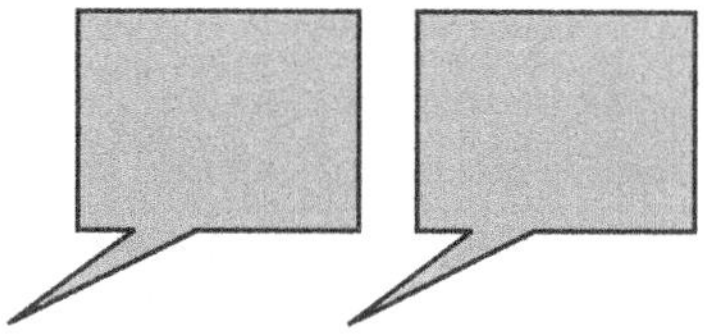

(B) 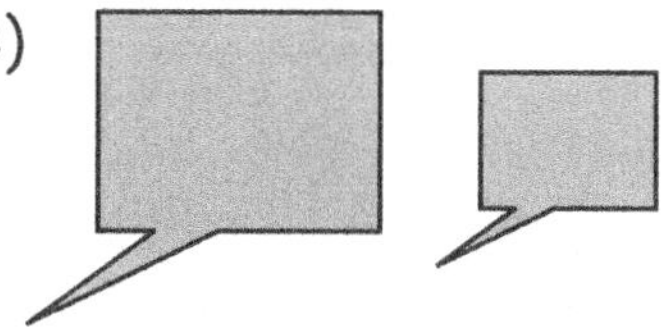

(C) 

(D) 

25. Below table shows the Zumba dance class practice. If the pattern shown in the table continues in the same way, how many minutes will Zumba practice class in 6th week ?

Zumba Class Duration	Time in Minutes
1	30
2	42
3	54
4	66

(A) 80 (B) 78

(C) 92 (D) 90

26. A box contains 15 red balls, 21 white balls, and 15 green balls. Liam picks up balls from the box. Which of the below describes the chances of picking a blue ball from the box ?

(A) Unlikely, but not impossible (B) Impossible

(C) Likely, but not certain (D) Certain

27. A box contains 5 white blocks, 6 black blocks and 12 silver blocks. Jason picks up a block from the box. What is the probability of picking a white block ?

(A) $\frac{12}{23}$ (B) $\frac{13}{23}$

(C) $\frac{5}{23}$ (D) $\frac{6}{25}$

28. What is the maximum number of line segments shown in the figure below ?

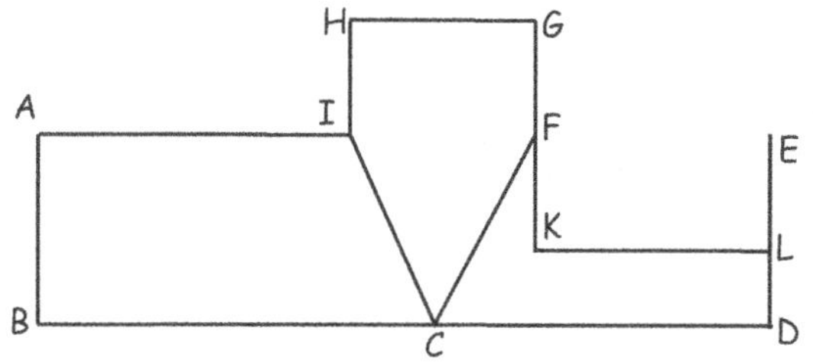

(A) 10

(B) 11

(C) 8

(D) 16

29. Which of the below is the correct statement for the Trapezoid MNOP

(A) Four right angles

(B) Opposite sides are parallel

(C) One pair of sides are parallel

(D) All sides are equal

30. Estimate the height of the tree ?

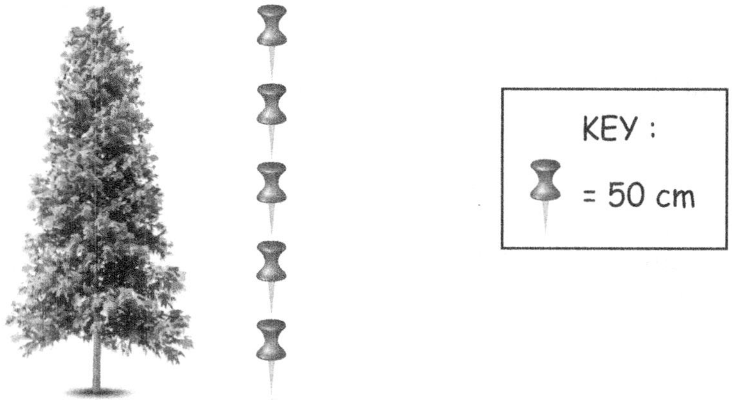

(A) 250 cm

(B) 200 cm

(C) 275 cm

(D) 320 cm

31. Find the height of the tear drop using the ruler provided ?

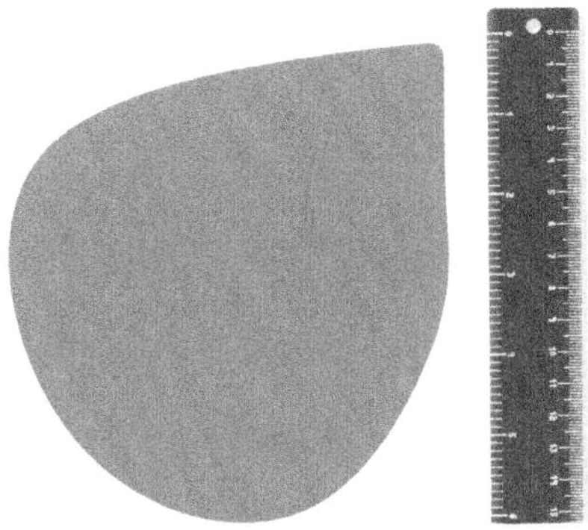

(A) 6 inches

(B) 5 inches

(C) 7 inches

(D) 6.9 inches

32. Which pair of the below lines are perpendicular ?

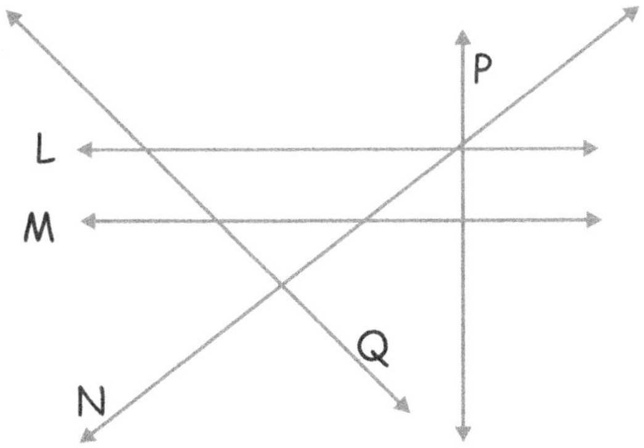

(A) L, M are perpendicular

(B) M, N are perpendicular

(C) L, N are perpendicular

(D) N, Q are perpendicular

33. Newly born Lions cub weighs 11 pounds. What is the cub's weight in ounces ?

(A) 175 ounces

(B) 177 ounces

(C) 186 ounces

(D) 176 ounces

34. Below model represents $\frac{5}{10}$

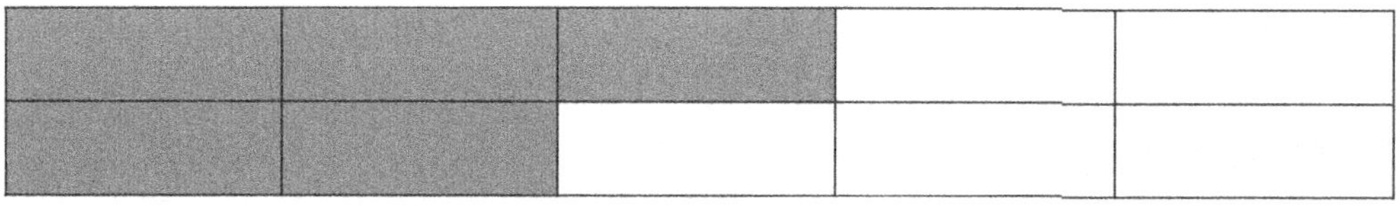

Which of the below is also equal to $\frac{5}{10}$?

(A) 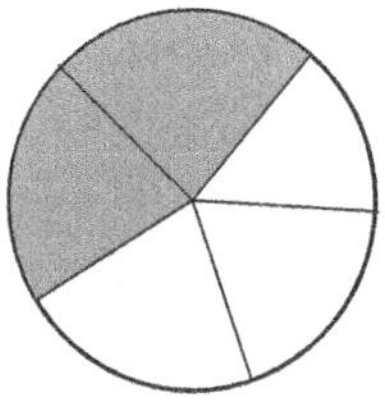

(B) 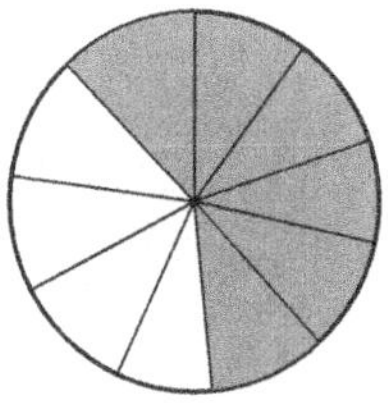

(C) 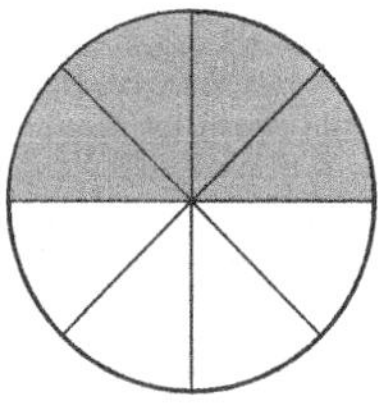

(D) 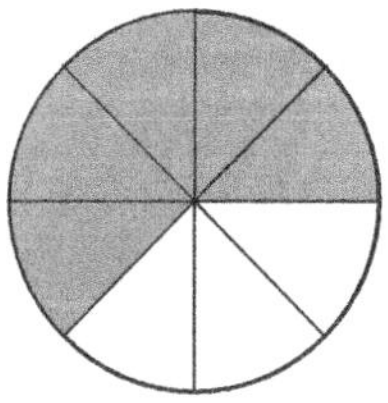

35. If below model equals to 1

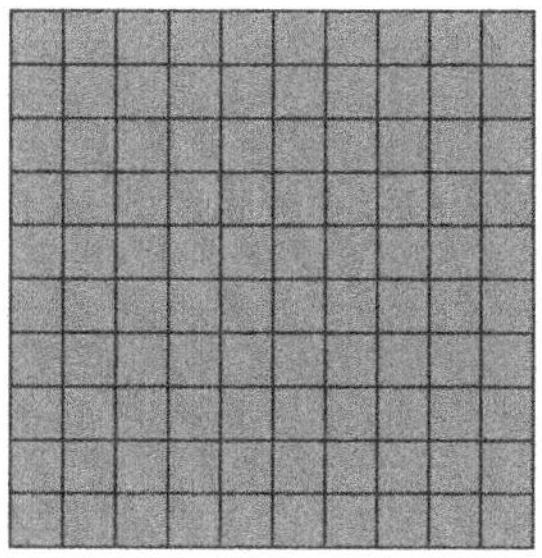

then

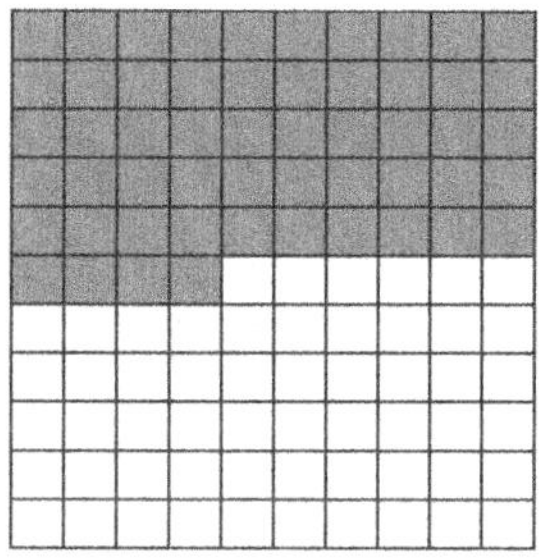

(A) 0.54

(B) 0.065

(C) 0.066

(D) 0.0066

36. Which of the below will round to 105.5 ?

(A) 105.155

(B) 105.586

(C) 104.355

(D) 105.533

37. Which of the below fractions has least value ?

(A) $\frac{2}{5}$ (B) $\frac{4}{8}$

(C) $\frac{1}{2}$ (D) $\frac{1}{3}$

38. Round 357,232 to the nearest ten thousand.

(A) 357,000 (B) 360,000

(C) 400,000 (D) 350,000

39. Round 807,076 to the nearest hundred.

(A) 807,000 (B) 807,080

(C) 807,100 (D) 807,076

40. Which of the below statement is correct ?

(A) $3203 < 2112$ (B) $999 < 998$

(C) $559 > 550$ (D) $4102 > 4109$

41. Write 70,707 in words.

(A) Seventy thousand, seven hundred and seven.
(B) Seventy thousand seven, seven hundred and seven.
(C) Seventy-seven thousand, seven hundred and seven.
(D) Seventy thousand seven, seventy-seven.

42. Which of the below statements is correct ?

(A) $\frac{5}{7} < \frac{7}{11}$

(B) $\frac{3}{9} > \frac{8}{11}$

(C) $\frac{3}{9} = \frac{7}{11}$

(D) $\frac{5}{7} > \frac{3}{5}$

43. Which ordered pair will represent the rectangle ?

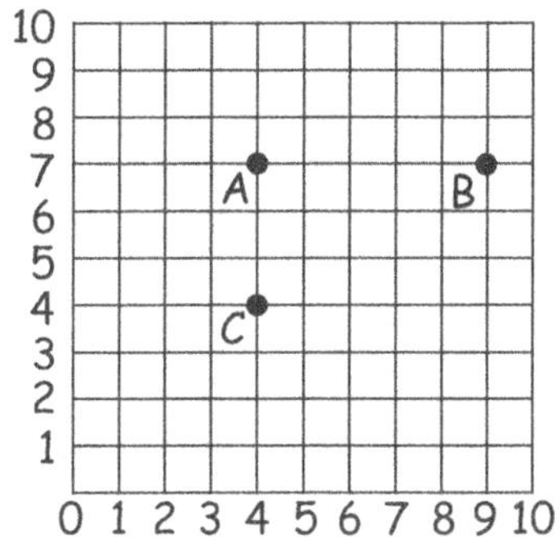

(A) (9,9)

(B) (9,4)

(C) (4,9)

(D) (4,4)

44. Dan measures strawberries on the balance scale. Based on the information given below estimate the weight of one strawberry.

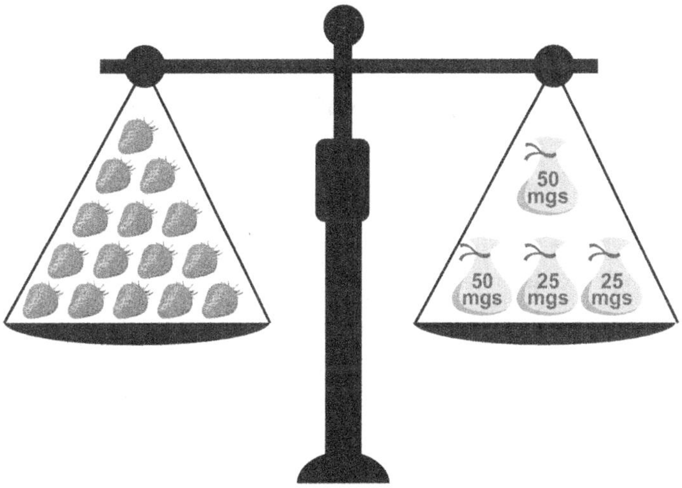

(A) 25 mgs

(B) 15 gms

(C) 30 mgs

(D) 10 mgs

Oak elementary school students participated in the favorite color survey.

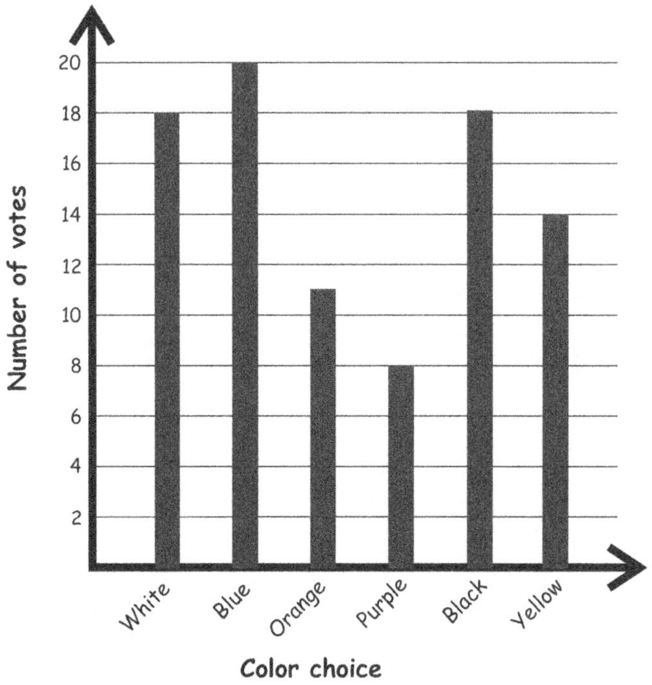

Based on the above data answer the below questions

45. How many students voted for blue and purple ?

(A) 30 (B) 25

(C) 28 (D) 37

46. How many students voted for orange and yellow ?

(A) 25 (B) 34

(C) 40 (D) 23

47. How many students participated in the survey ?

(A) 54 (B) 76

(C) 97 (D) 89

48. Which of the below are congruent ?

(A) 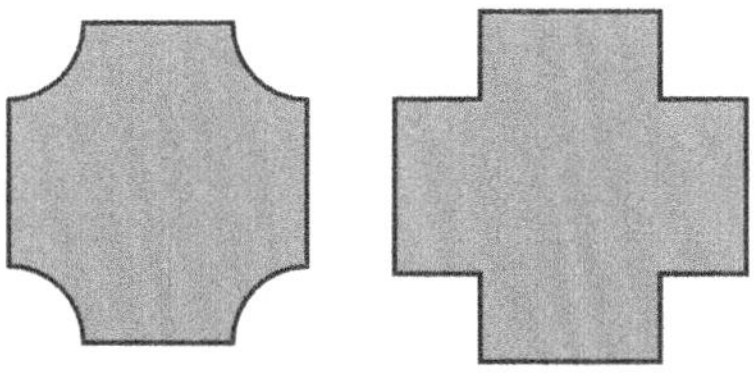

(B) 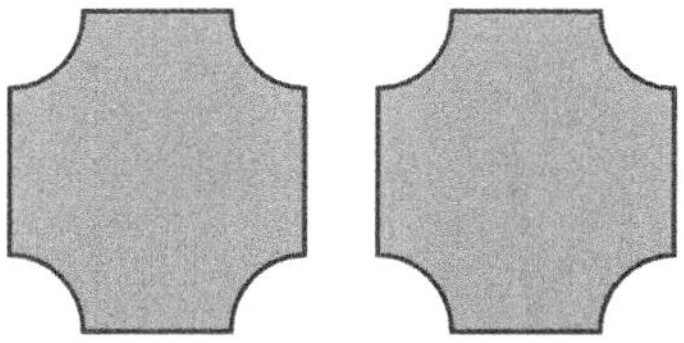

(C) 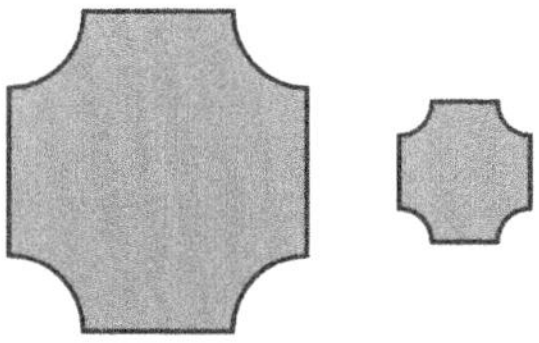

(D) 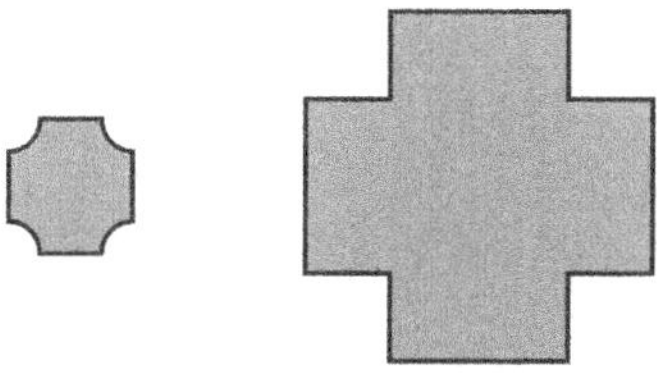

49. Which of the below option is an example for reflection ?

(A) 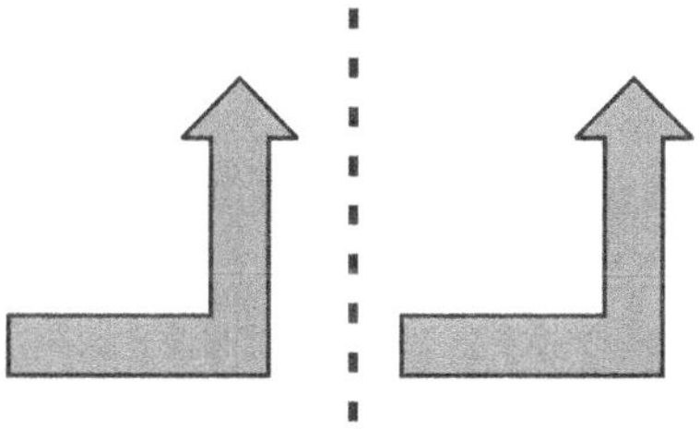

(B) 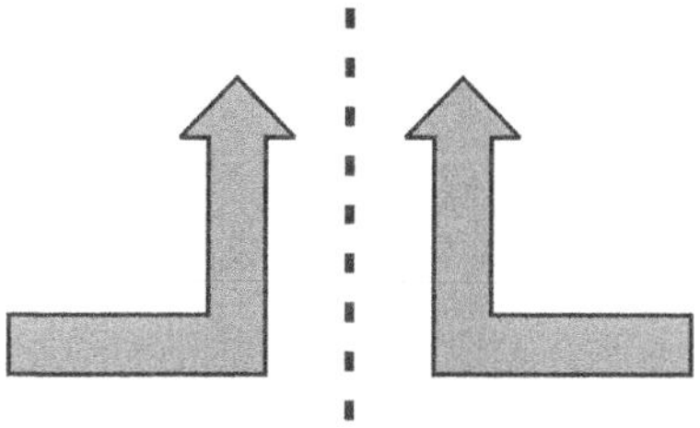

(C) 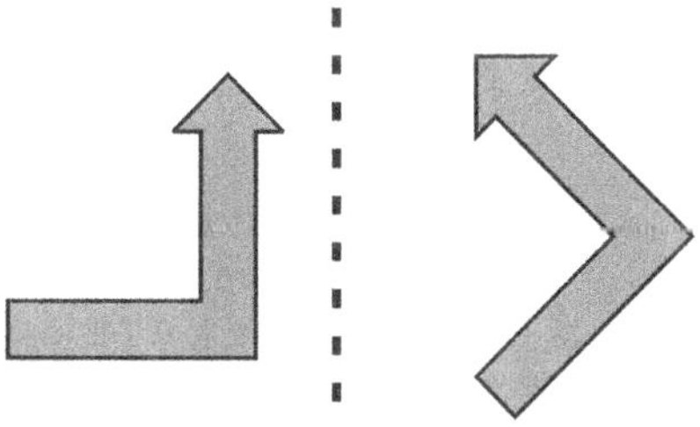

(D) 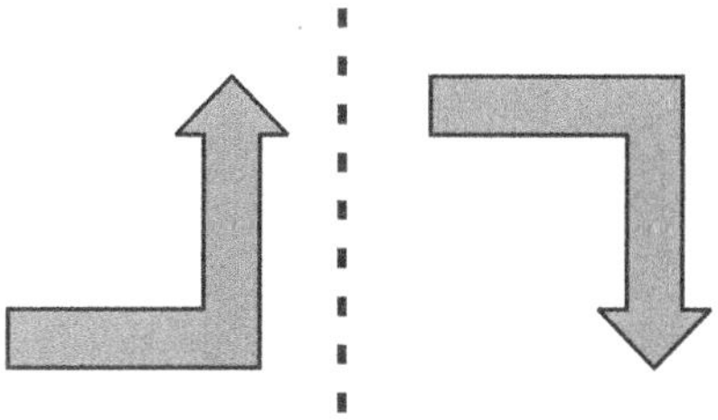

50. Find the perimeter of the below figure

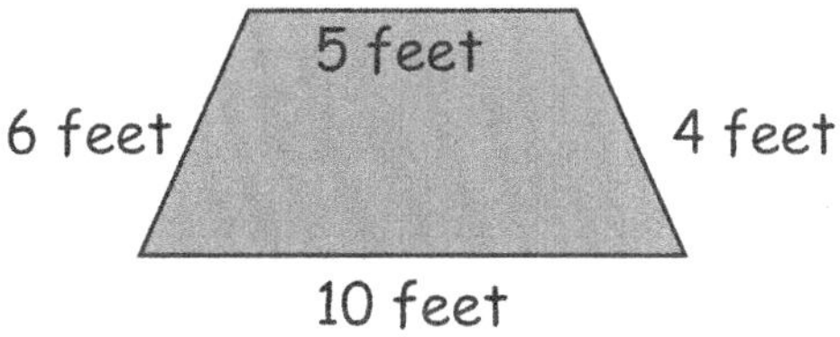

(A) 25 feet

(B) 10 feet

(C) 20 feet

(D) 15 feet

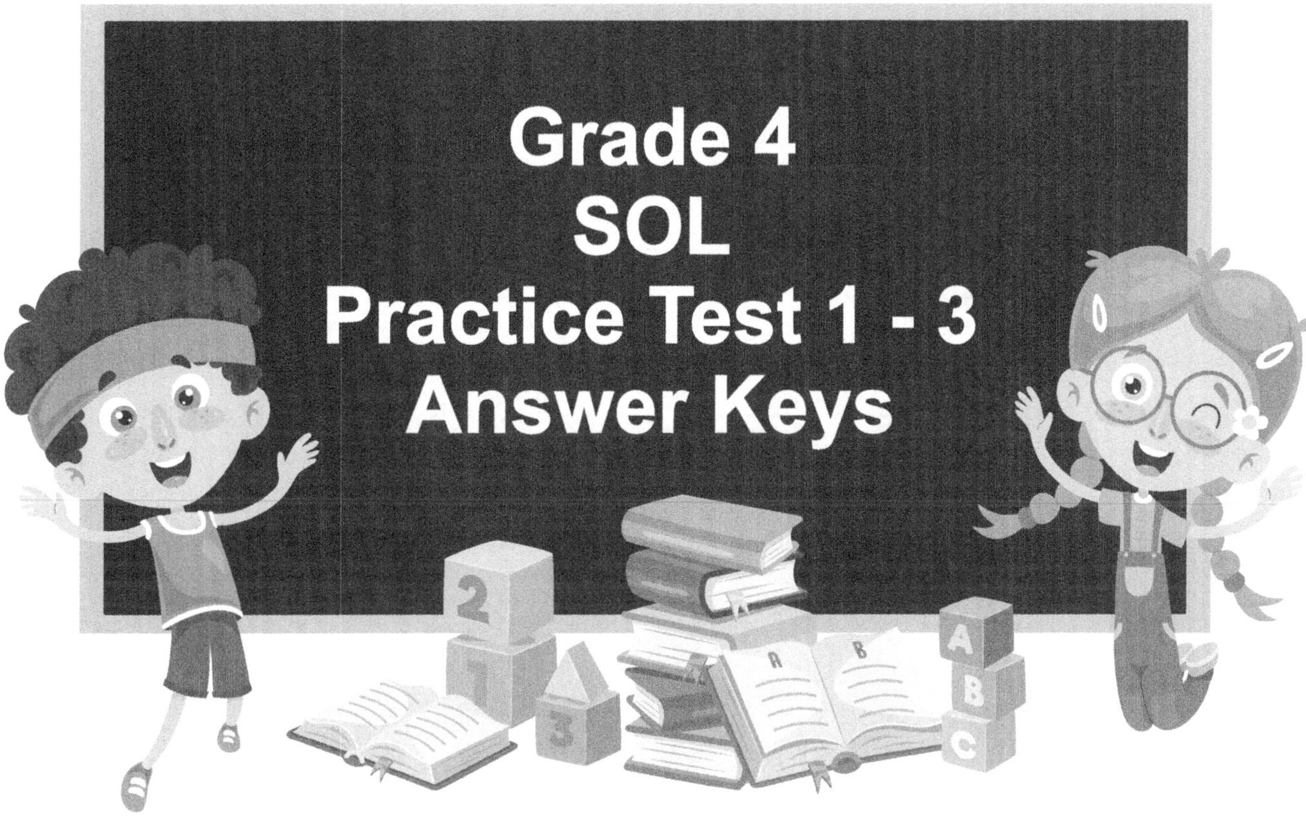
Grade 4
SOL
Practice Test 1 - 3
Answer Keys

Answer Key Test - 1

1. A
2. A
3. B
4. A
5. D
6. C
7. D
8. D
9. B
10. A
11. D
12. B
13. D
14. C
15. B

16. C

17. D

18. A

19. C

20. A

21. A

22. C

23. A

24. B

25. A

26. C

27. D

28. B

29. D

30. A

31. D

32. B

33. A

34. C

35. A

36. C

37. D

38. D

39. A

40. D

41. B

42. A

43. B

44. C

45. C

46. D

47. A

48. B

49. D

50. C

Answer Key Test - 2

1. B
2. D
3. A
4. C
5. C
6. B
7. D
8. C
9. C
10. A
11. A
12. D
13. B
14. A
15. C

16. A

17. C

18. D

19. B

20. D

21. A

22. A

23. D

24. A

25. B

26. B

27. D

28. C

29. A

30. C

31. B

32. D

33. C

34. D

35. C

36. A

37. A

38. A

39. C

40. A

41. D

42. B

43. C

44. A

45. D

46. C

47. D

48. B

49. A

50. D

Answer Key Test - 3

1. C
2. D
3. C
4. C
5. D
6. D
7. A
8. B
9. A
10. B
11. A
12. D
13. C
14. D
15. A

16. B

17. A

18. D

19. D

20. B

21. C

22. C

23. A

24. A

25. D

26. B

27. C

28. D

29. C

30. A

31. A

32. D

33. D

34. C

35. A

36. D

37. D

38. B

39. C

40. C

41. A

42. D

43. B

44. D

45. C

46. A

47. D

48. B

49. B

50. A

GRADE 4
SOL
Score Calculation
2
1
3
A
B
C

Score calculation

If you get this many times correct :	Then your converted scale scor is :
0	000
1	217
2	249
3	268
4	282
5	293
6	303
7	311
8	319
9	325
10	331
11	337
12	343
13	348
14	353
15	357
16	362
17	366
18	371
19	375

20	379
21	383
22	387
23	391
24	395
25	399
26	403
27	407
28	411
29	415
30	419
31	423
32	427
33	432
34	436
35	441
36	445
37	450
38	456
39	461

SCORE

40	467
41	473
42	480
43	487
44	496
45	505
46	516
47	531
48	550
49	582
50	600

Made in the USA
Monee, IL
18 October 2026

66036199R00052